# Artificial Intelligence and Machine Learning in the Thermal Spray Industry

This book details the emerging area of the induction of expert systems in thermal spray technology, replacing traditional parametric optimization methods like numerical modeling and simulation. It promotes, enlightens, and hastens the digital transformation of the surface engineering industry by discussing the contribution of expert systems like Machine Learning (ML) and Artificial Intelligence (AI) toward achieving durable Thermal Spray (TS) coatings.

*Artificial Intelligence and Machine Learning in the Thermal Spray Industry: Practices, Implementation, and Challenges* highlights how AI and ML techniques are used in the TS industry. It sheds light on AI's versatility, revealing its applicability in solving problems related to conventional simulation and numeric modeling techniques. This book combines automated technologies with expert machines to show several advantages, including decreased error and greater accuracy in judgment, and prediction, enhanced efficiency, reduced time consumption, and lower costs. Specific barriers preventing AI's successful implementation in the TS industry are also discussed. This book also looks at how training and validating more models with microstructural features of deposited coating will be the center point to grooming this technology in the future. Lastly, this book thoroughly analyzes the digital technologies available for modeling and achieving high-performance coatings, including giving AI-related models like Artificial Neural Networks (ANN) and Convolutional Neural Networks (CNN) more attention.

This reference book is directed toward professors, students, practitioners, and researchers of higher education institutions working in the fields that deal with the application of AI and ML technology.

# Multi-Scale and Multi-Functional Materials: Design, Development, and Applications

*Series Editors:*
Chander Prakash and Alokesh Pramanik

Artificial Intelligence and Machine Learning
in the Thermal Spray Industry
Practices, Implementation, and Challenges
*Lalit Thakur, Hitesh Vasudev, Jashanpreet Singh and Gaurav Prashar*

# Artificial Intelligence and Machine Learning in the Thermal Spray Industry

## Practices, Implementation, and Challenges

Lalit Thakur, Hitesh Vasudev,
Jashanpreet Singh, and Gaurav Prashar

CRC Press
Taylor & Francis Group
Boca Raton London New York

CRC Press is an imprint of the
Taylor & Francis Group, an **informa** business

Designed cover image: Shutterstock, Jashanpreet Singh

MATLAB® is a trademark of The MathWorks, Inc. and is used with permission. The MathWorks does not warrant the accuracy of the text or exercises in this book. This book's use or discussion of MATLAB® software or related products does not constitute endorsement or sponsorship by The MathWorks of a particular pedagogical approach or particular use of the MATLAB® software.

First edition published 2024
by CRC Press
2385 NW Executive Center Drive, Suite 320, Boca Raton FL 33431

and by CRC Press
4 Park Square, Milton Park, Abingdon, Oxon, OX14 4RN

CRC Press is an imprint of Taylor & Francis Group, LLC

© 2024 Lalit Thakur, Hitesh Vasudev, Jashanpreet Singh, and Gaurav Prashar

*Library of Congress Cataloging-in-Publication Data*
Names: Thakur, Lalit, author. | Vasudev, Hitesh, author. |
Singh, Jashanpreet, author. | Prashar, Gaurav, author.
Title: Artificial intelligence and machine learning in the thermal spray
industry : practices, implementation, and challenges / Lalit Thakur,
Hitesh Vasudev, Jashanpreet Singh, Gaurav Prashar.
Description: Boca Raton : CRC Press, 2024. | Series: Multi-Scale and
Multi-Functional Materials : Design, Development, and Applications. |
Includes bibliographical references and index.
Identifiers: LCCN 2023025073 (print) | LCCN 2023025074 (ebook) |
ISBN 9781032502243 (hardback) | ISBN 9781032510040 (paperback) |
ISBN 9781003400660 (ebook)
Subjects: LCSH: Metal spraying. | Artificial intelligence—Industrial applications.
Classification: LCC TS655 .T465 2024 (print) | LCC TS655 (ebook) |
DDC 671.7/34028653—dc23/eng/20230821
LC record available at https://lccn.loc.gov/2023025073
LC ebook record available at https://lccn.loc.gov/2023025074

ISBN: 978-1-032-50224-3 (hbk)
ISBN: 978-1-032-51004-0 (pbk)
ISBN: 978-1-003-40066-0 (ebk)

DOI: 10.1201/9781003400660

Typeset in Times
by codeMantra

# Contents

# About the Authors

**Dr. Lalit Thakur** is currently working as an assistant professor in the Department of Mechanical Engineering at the National Institute of Technology (NIT) in Kurukshetra, India. He has obtained his M.Tech. (Welding Engineering) and Ph.D. degrees from the Indian Institute of Technology (IIT) Roorkee, India. For the last 13 years, he has been continuously exploring new possibilities in Welding Engineering and Thermal Spray Technology. He has guided many Ph.D. and Master candidates in the area of Welding and Thermal Spraying. He has authored more than 80 research publications in various international journals, books, and conferences of repute.

**Dr. Hitesh Vasudev** is working as a full-time professor in the School of Mechanical Engineering and Division of Research and Development at Lovely Professional University, Jalandhar, India. He has received a Ph.D. in Mechanical Engineering from Guru Nanak Dev Engineering College, India. His areas of research are surface engineering, bimodal coatings, and additive manufacturing. He has more than 10 years of teaching and research experience. He has authored more than 100 research papers and 3 books, and has organized international conferences.

**Dr. Jashanpreet Singh** is currently working at the University Centre of Research and Development, Chandigarh University as an assistant professor. He obtained his Ph.D. degree from the Thapar Institute of Engineering and Technology, Patiala in 2019. He has authored more than 60 research publications in various international journals (SCIE/Scopus), books, and conferences of repute. He has a teaching experience of more than three years and industrial experience of more than two years. His areas of research are tribology-solid particle erosion, thermal spray coatings, CFD simulation, artificial intelligence, machine learning and regression tools, process optimization, composites, and materials characterization.

**Dr. Gaurav Prashar** is an associate professor and head of the Department of Mechanical Engineering at the Rayat Bahra Institute of Engineering and Nano-Technology, Hoshiarpur, India. His areas of research are surface engineering and additive manufacturing. He has more than 15 years of teaching and research experience. He has authored more than 20 research papers (SCI/Scopus). The research outcomes have been published in reputed journals (SCI/Scopus) such as the *Journal of Cleaner Production, Surface and Coatings Technology, Journal of Thermal Spray Technology, Engineering Failure Analysis*, and *Surface Topography: Metrology and Properties*.

# 1 Artificial Intelligence in Thermal Spray Industry
## *Introduction and Benefits*

## ABBREVIATIONS

| | |
|---|---|
| **AI** | Artificial intelligence |
| **ANN** | Artificial neural networks |
| **CS** | Cold spray |
| **HVOF** | High-velocity oxy-fuel |
| **ML** | Machine learning |
| **R&D** | Research and development |
| **SVM** | Support vector machine |
| **TS** | Thermal spraying |
| **PS** | Plasma spray |

## 1.1 INTRODUCTION: ARTIFICIAL INTELLIGENCE IN THERMAL SPRAY COATINGS

Artificial intelligence (AI) using well-known machine learning (ML) techniques from the computer science field is broadly affecting many aspects of Industrial Revolution 4.0, including science and technology, the manufacturing industry, and even our day-to-day lives. The ML methods have been designed to analyze a large amount of data to gain insightful information, classify, predict, and make judgments based on evidence in unique ways. This will encourage the development of novel applications that support AI's sustained growth in modern-day industries [1,2]. When it comes to physics-based modeling, ML provides a new approach to leverageable datasets and data-driven methodologies. The creation of these new spectacles is a major success for AI and ML methods. ML is helpful because it can analyze huge data sets and figure out how they are set up in a way that humans can't [3]. The commonalities between different branches of AI and their scope are shown schematically in Figure 1.1 [4].

Data mining (like Convolutional Neural Networks) and Knowledge Discovery in Databases techniques are analogous to ML. So are other branches of statistics (like regression models) and pattern recognition (like failure analysis). Figure 1.2 is a simplified illustration of a common ML implementation in the field of TS. To use ML methodology, one must first create a dataset through real experimentation, simulation, or mining. Each variable, such as composition, density, and hardness, is a separate input in the dataset. Therefore, information on the composition of glass, its method of manufacture, and other such aspects may be obtained from a database

DOI: 10.1201/9781003400660-1

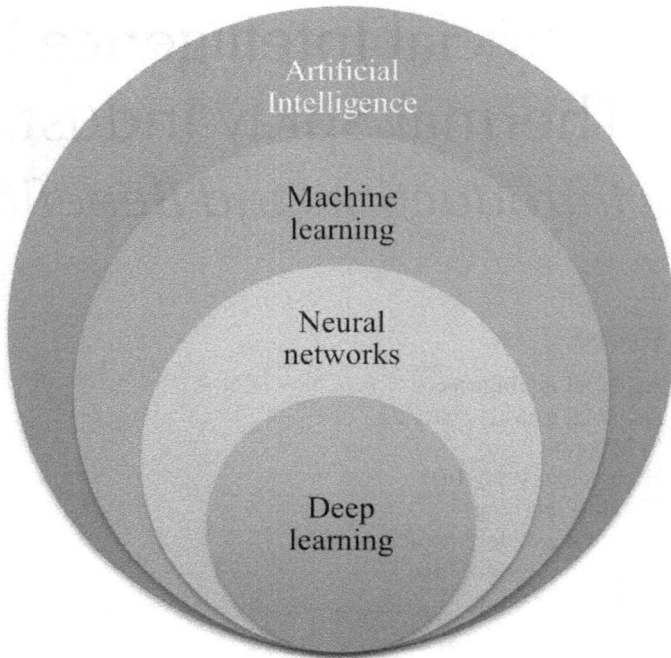

**FIGURE 1.1**  Illustration demonstrating the general scope of artificial intelligence.

holding attributes (components or substances). ML is used to construct a prediction model after correlations have been discovered within the dataset [3]. There are two main categories of ML algorithms: supervised and unsupervised. In the following sections, we'll dive further into the specifics of supervised and unsupervised ML methods. Unsupervised learning makes use of clustering models for prediction, while supervised learning uses regression and classification models. To reliably anticipate outcomes using AI and ML methods, it is necessary to adhere to a variety of best practices. The functional processes of ML applications include issue formulation, data collection, model and loss function tuning, data partitioning, and under-/over-fitting. The next sections provide a more in-depth explanation of these procedural stages. As can be seen in Figure 1.3, ML methods may be broadly classified into three broad categories. The ML algorithm with the lowest percentage of error is chosen as the best suited or fitted. Underfitting and overfitting are prevented while the ML algorithms are verified. These techniques are well discussed in Chapter 2 with details.

## 1.2  ARTIFICIAL INTELLIGENCE BASICS AND ITS HISTORY

A 60-year-old field known as AI is an organization of methods and concepts that includes mathematical logic, computer science, probability, and computational neuroscience. The field of computer science known as AI is constantly developing and is focused on creating software that can conduct complex and intelligent calculations comparable to those performed regularly. The term AI refers to techniques, tools, and systems that simulate human techniques for learning and problem-solving

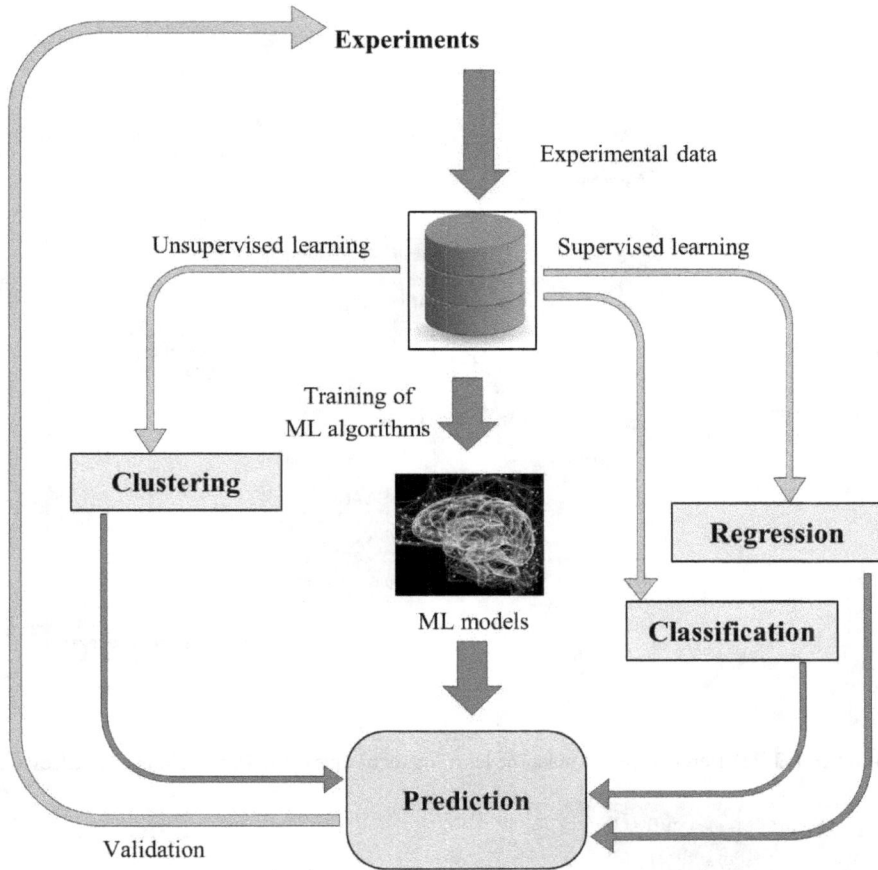

**FIGURE 1.2** Schematic sketch of machine learning applications [13]. Permissions from Elsevier.

using brain activity [5]. The brief history of AI is represented in Figure 1.4. John McCarthy, who invented the phrase AI, during a symposium at Dartmouth College in 1956, is credited with starting current AI research [6]. This represented the beginning of the AI science discipline, and within this era, the focus was mainly on naive algorithms. However, this algorithm is suitable only for small texts. The advancement in the years that followed after 1960 was astounding. Many scientists and researchers concentrated on automated reasoning and employed AI for algebraic problem-solving and the proof of mathematical theorems. These achievements gave many AI pioneers unbridled hope and supported their conviction that fully smart machines would be created soon. They quickly learned, though, that a long way had to go before machines exhibited intelligence on par with that of humans. The logic-based programs were unable to solve many nontrivial tasks. The availability of computational resources to solve ever-more complex issues was another difficulty. As a result, organizations and investors ceased to support these AI projects that fell short of expectations.

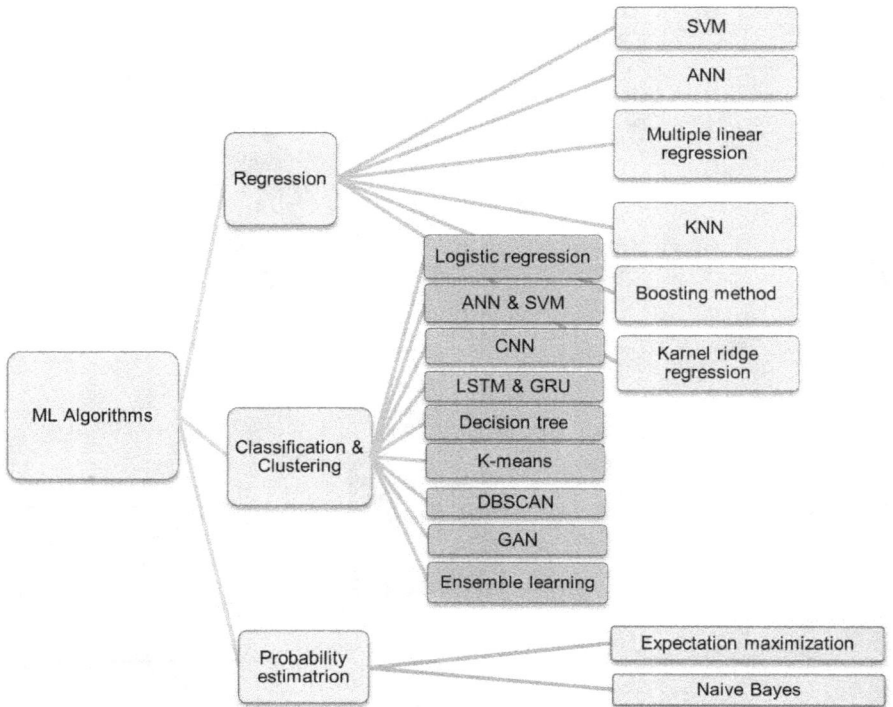

**FIGURE 1.3** Different types of machine learning techniques [13]. Permissions from Elsevier.

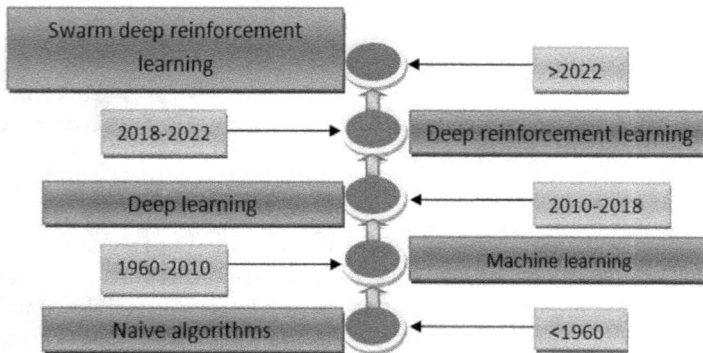

John McCarthy, Father of modern day AI

**FIGURE 1.4** Schematic illustration of flow chart representing the history of artificial intelligence.

In the 1980s, various academic and research institutes developed a form of AI system that summarizes several fundamental principles from expert knowledge to assist nonexperts in making certain decisions, which were known as "Expert systems." Examples include the MYCIN created by Stanford University, and the XCON, created by Carnegie Mellon University. For the first time, an expert system used logic rules generated from expert knowledge to address issues in the real world. The understanding that made machines "smarter" served as the foundation of AI research throughout this time. The expert system did, however, gradually point out many drawbacks, including privacy technologies, a lack of adaptability, low versatility, and high maintenance costs.

In the meantime, the fifth-generation computer project, which received significant funding from the Japanese government, fell short of its initial objectives. The funding for AI development was cut off once more, and the field was in its second-lowest position ever. Geoffrey Hinton and colleagues [7,8] significantly contributed to AI in 2006 by suggesting a method for developing DNN. Due to this, DL algorithms have emerged as one of the most active areas in AI research. DL is a subtype of ML that uses representation learning and several layers of neural networks [9]. Conversely, ML is a component of AI that enables a computer or program to learn and develop intelligence without the need for human interaction. A rising number of innovative neural network architectures and training techniques have been developed to enhance the representational learning capability of DL and broaden it into more general applications. High-throughput data can be analyzed using ML techniques to classify, forecast, and make novel decisions based on evidence. Hence, in every sphere of life, AI technologies have had tremendous success. They have also demonstrated their worth as the foundation of scientific thinking and practical applications. Conversely, ML is a component of AI that enables a computer or program to learn and develop intelligence without the need for human interaction. To enhance the representational learning capability of DL and broaden it into more general applications, a rising number of innovative neural network architectures and training techniques have been developed. High-throughput data can be analyzed using ML techniques to classify, forecast, and make novel decisions based on evidence. Hence in every sphere of life, AI technologies have had tremendous success. They have also demonstrated their worth as the foundation of scientific thinking and practical applications.

Although there are many different aspects of AI like reactive machines AI, self-aware AI, limited memory AI, and the theory of mind AI. However, perceptual, cognitive, and decision-making intelligences are all involved in the development phase of AI. There are primarily two types of developments in the field of AI. First, we find methods and software like expert systems, which attempt to mimic human cognition by deducing results from a predetermined set of rules. The second category consists of systems that simulate how the brain functions, such as artificial neural networks (ANNs) [5]. ANNs were initially established by two researchers named Warren McCulloch and Walter Pitts in 1943, who created a computational model mainly for neural networks that was based on threshold logic techniques [10]. This paradigm shift cleared the way for research to be divided into two streams, one concentrating on biological processes and the other focusing on using neural networks in AI.

Cortes and Vapnik created the support vector machine, a very intelligent sort of perceptron, during the second period of inactivity [11]. It continued and eventually surpassed neural networks. Since a support vector machine performs better with less computational time and training, many researchers have focused on studying it rather than neural networks with several adaptive hidden layers. Einerson et al. [12] introduced the initial concept of applying neural computing to TS processes in 1993. The relationship between specific processing parameters and the in-flight particle characteristics using an ANN was explained in that study. The following benefits were provided by ANN structures during this study:

- If the responses and parameters are quantified, they represent any input-output relationship;
- Finding nonlinear and complex correlations that are encoded in the ANN structure;
- There is no need to make assumptions beforehand;
- Incorporating the variability and fluctuations associated with the experimental sets;
- Restricting the number of experiments;
- The discovery of new correlations is made possible by adding more experimental sets, called a progressive system (also known as "continuous" learning).

However, there are some negative aspects as well, such as:

- The requirement for a database that can be created by taking process history into account;
- A physical interpretation was absent;
- Understanding the system parameters is also crucial before the implementation of the ANN.

## 1.3   BASICS OF ANN MODELS

ANN is a model of computational neurons designed to mimic human brain activity during learning. Edges link neurons to one another. As learning progresses, new relative importance is placed on these neurons and edges. In Figure 1.5a, we see a basic ANN with only one hidden layer. Input, hidden, and output layers are all present. Ultimately, the obtained output is compared to the real production, and weights are modified depending on the difference. On the other hand, ANNs with numerous layers are used to address a wide variety of issues. The complexity of the challenges arises as the number of hidden layers grows. This network structure is employed because it simplifies otherwise difficult challenges. A multilayered ANN is shown in Figure 1.5b.

## 1.4   ML IN THERMAL SPRAY INDUSTRY

The TS technique has undergone substantial development and is currently utilized widely across all significant engineering fields. A high-quality and high-performance TS coating meets the needs of diverse industrial applications. It mainly depends on the states of in-flight particles, which are mostly controlled by process parameters

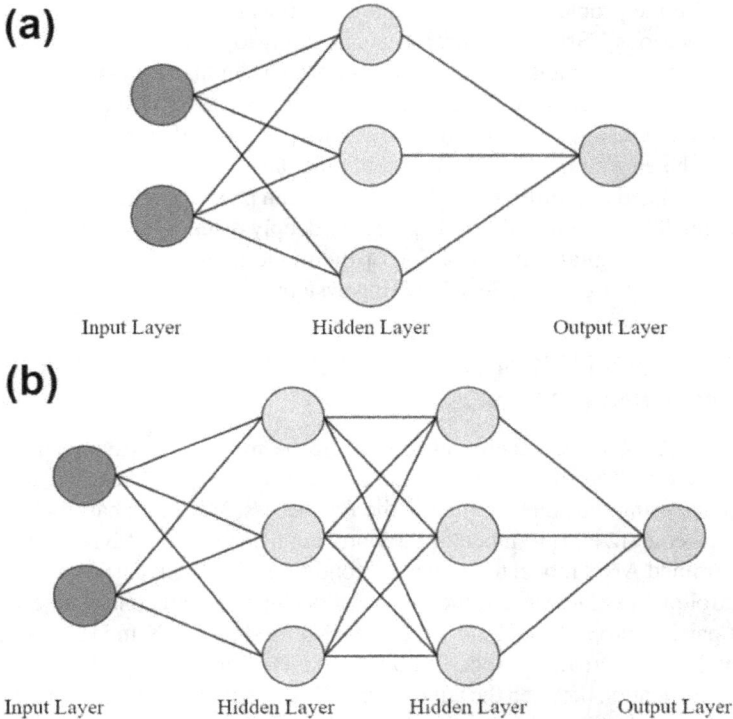

**FIGURE 1.5** (a) Single and (b) multiple hidden layers in an artificial neural network model [13]. Permissions from Elsevier.

and impact the coating qualities in the TS spray process. Many chemical and thermodynamic processes take place throughout the deposition process, and obtaining a thorough multiphysical model is difficult for any TS process. The most popular approach is to determine the effect of TS process parameters using Taguchi's design of experiments approach, which is a preliminary method to determine the optimal process parameters and separate the important and minor factors [14–16]. However, since the coating quality is typically impacted by a combined effect of various process parameters, it may fail to attain the precise ideal parameters. The spray processes have also been extensively simulated and controlled using numerical modeling and simulation [17–19], which frequently give greater attention to examining the evolution of the spray and combustion processes. However, numerical modeling struggles to simulate the actual behaviors because of the complex multiphysical phenomena of the TS process. The TS coatings must therefore be carefully analyzed, predicted, and optimized using a proper approach.

TS has developed as a highly automated spraying process during the last decade due to industrial automation, but contemporary TS applications also need to be intelligent. Determining the relationship between operating parameters, in-flight particle behavior, and coating characteristics via a central automated system is the need of the hour to achieve durable and quality TS coatings. This necessity has inspired researchers to create a reliable methodology that uses modern digital techniques like

AI to resolve the problems of TS-deposited coatings under various operating conditions. The various TS spray control systems can also be configured to incorporate models, creating an intelligent control system. The industry must develop greater intelligence, connectivity, and the ability to integrate easily into spray systems and current manufacturing environments. New computing-based approaches are required to process the enormous amount of quantifiable data and use it to control TS operations for enhanced reliability and robustness. AI can play a vital role in achieving this requirement. The purpose of this chapter is to deeply discuss AI, and its benefits that can promote the digital transformation of the modern surface engineering industry and help in achieving a durable TS coating system.

## 1.5   IMPLEMENTATION OF ANN STRUCTURES FOR PLASMA SPRAY (PS) COATINGS

Researchers [20–23] examined and forecasted the microstructure and properties of coatings, such as porosity percentage, the content of unmelted particles, and various components, using the ANNs to model the PS process. Modular analysis was used by Choudhury et al. [24,25] to model the PS process using several ANNs, and they then used the trained ANN model to predict the condition of in-flight particles based on various control process factors. To study the relationship between control parameters and in-flight particle properties, Kanta et al. [26–28] used an ANN model to analyze the PS method used to create the specific coating. ANNs were used by Zhang et al. [29] to develop a correlation between the atmospheric plasma spraying process and the features of in-flight particles as well as the impact of particle temperature on the microstructure of coatings. Through a neural network, Liu et al. [30] investigated the link between spray gun power, particle temperature, and particle velocity. The effect of the plasma and powder injection parameters on the in-flight particle attributes (average velocity, particle temperature, and particle diameter) for an $Al_2O_3$ feedstock with 13 wt% $TiO_2$ was examined in one of the earliest uses of ML in TS. The predicted in-flight characteristics using the ML approach were in agreement with the measured values after performing a validation phase, a training step, and a test step [31]. Additionally, coating structural properties were incorporated into the ANN in a subsequent investigation. Therefore, this model was employed to examine the opposing trends of the deposition yield and coating porosity [26], as well as the particle melting status [32].

## 1.6   IMPLEMENTATION OF ANN STRUCTURES FOR HVOF SPRAY COATINGS

The above-mentioned experimental studies were concentrated on the plasma spraying method. ANN model was also employed for the high-velocity oxy-fuel (HVOF) process. Only a limited number of studies examine the HVOF spray procedure using the ANN model. To investigate the magnetic characteristics of an HVOF-sprayed FeNb alloy, Cherigui et al. [33] developed two ANN models: Model A was used to link system parameters to features of the microstructure, whereas Model B was used to link system parameters to the characteristics of magnetism. Zhang et al. [34] used an ANN model to predict the structural characteristics (porosity and hardness) of the HVOF-sprayed

NiCrAlY coating in terms of the relationship between the system parameters (oxygen/ fuel gas stoichiometric ratio and stand-off distance). Using an ANN model, Kamnis et al. [35] conducted an intriguing investigation on airborne acoustic emission during the HVOF spray process to highlight the significant impact of the spray distance and powder feed rate on the micro-hardness of coatings. In studies conducted by Mojena et al. [36], an ANN model has also been used to forecast the erosive wear rate for WC-CoCr coatings that are coated by flame spray and HVOF. The results of this investigation suggest that porosity, followed by a combination of micro-hardness and fracture toughness, exerts the largest influence over the rate of erosion.

The HVOF spray process has been thoroughly studied using the ANN model by Liu et al. [37], which has been used to link the process parameters with the mechanical performance of coatings. The findings indicate that the parameters for micro-hardness and porosity should be considered in the following order: spray distance, oxygen flow rate, and $CH_4$ flow rate, and for wear rate, the oxygen flow rate, stand-off distance, and $CH_4$ flow rate should be considered. However, in this study, the velocity and temperature of in-flight particles, which are intermediate factors, have not been taken into account.

The impacts on in-flight particle characteristics and microstructural coating qualities of spraying $Cr_3C_2$_25NiCr HVOF feedstocks were investigated again by Liu et al. [38]. In Figure 1.6, the framework is shown. Two models were selected: for Model 1, stand-off distance, gas flow rate, and fuel flow rate were chosen as inputs, whereas temperature and velocity of particles were selected as targets. In Model 2,

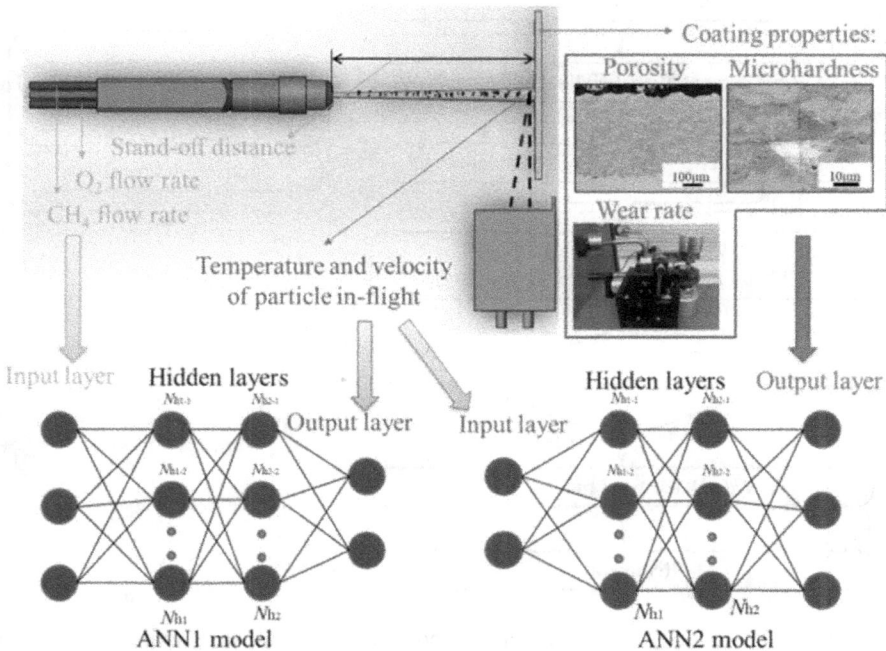

**FIGURE 1.6** High-velocity oxy-fuel process representation using artificial neural networks, along with the performance of the resultant coating [38]. Permissions from Springer Nature.

the temperature and velocity of particles were selected as inputs, while coating properties were targets. Other TS approaches will also be encouraged to use this cutting-edge ANN approach, leading to improved coating performance controls in the near future.

## 1.7 IMPLEMENTATION OF ANN STRUCTURES FOR COLD SPRAY (CS) SPRAYED COATINGS

Repairing and additive manufacturing are two areas where cold spraying may find use. In preparation for additive manufacturing, ANN was to forecast the coating thickness profile of CS multilayered Cu coatings on both flat and curved specimens [39]. The critical velocity of the CS is a crucial component that determines the adherence of particles during the CS process. ANN was also used to study how basic feedstock properties affect the critical velocity in CGS [40]. The ANN method's flowchart for determining the critical velocity is shown in Figure 1.7. The findings for the majority of the materials under investigation were more consistent than those produced using an empirical method that had already been published. The critical velocity was shown to be most affected by mechanical material factors (tensile and yield strength) compared to thermal parameters (melting temperature and thermal conductivity).

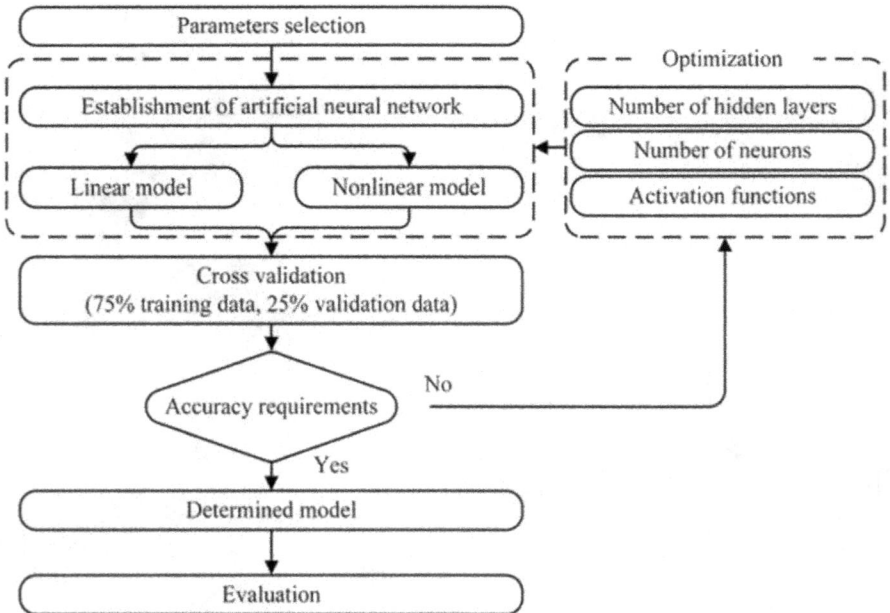

**FIGURE 1.7** Artificial neural network method's flowchart for determining the critical velocity [39]. Permissions from Springer Nature.

## 1.8 SCOPE AND CONCLUSIONS

The authors have thoroughly analyzed the AI technologies and their scope in TS for achieving high-performance coatings. AI-related benefits were also discussed. The key outcomes are as follows:

- It is anticipated that with suitable models and ML technologies, it will be possible to create coatings with certain target properties using fewer tests.
- Only a small amount of research examines the HVOF and CS spray procedures using the ANN model. Therefore, other TS approaches will also be encouraged to use this cutting-edge ANN approach, leading to improved coating performance controls.
- A research area that has drawn a lot of interest is the study of DL-based data augmentations. When only a few images are available, data augmentation is a particularly effective method for TBC microstructure applications.
- In a nutshell, much more work needs to be done in the upcoming years to identify the data sets and ML techniques that are necessary for putting into practice reliable control strategies for creating the best coatings and managing TS operations in industrial production.

## REFERENCES

[1] Singh J, Singh S. Neural network prediction of slurry erosion of heavy-duty pump impeller/casing materials 18Cr-8Ni, 16Cr-10Ni-2Mo, super duplex 24Cr-6Ni-3Mo-N, and grey cast iron. *Wear* [Internet]. 2021;476:203741. Available from: https://doi.org/10.1016/j.wear.2021.203741.

[2] Singh J. Analysis on suitability of HVOF sprayed Ni-20Al, Ni-20Cr and Al-20Ti coatings in coal-ash slurry conditions using artificial neural network model. *Ind Lubr Tribol*. 2019;71:972–982.

[3] Bulgarevich DS, Tsukamoto S, Kasuya T, et al. Pattern recognition with machine learning on optical microscopy images of typical metallurgical microstructures. *Sci Rep* [Internet]. 2018;8:3–9. Available from: https://doi.org/10.1038/s41598-018-20438-6.

[4] Shobha G, Rangaswamy S. *Machine Learning [Internet]*. 1st ed. Handb. Amsterdam: Stat. Elsevier B.V.; 2018. Available from: https://doi.org/10.1016/bs.host.2018.07.004.

[5] Agatonovic-Kustrin S, Beresford R. Basic concepts of artificial neural network (ANN) modeling and its application in pharmaceutical research. *J Pharm Biomed Anal*. 2000; 22:717–727.

[6] Xu Y, Liu X, Cao X, et al. Artificial intelligence: A powerful paradigm for scientific research. *The Innovation*. 2021;2:100179.

[7] Hinton GE, Osindero S, Teh Y-W. A fast learning algorithm for deep belief nets. *Neural Comput*. 2006;18:1527–1554.

[8] Hinton GE, Salakhutdinov RR. Reducing the dimensionality of data with neural networks. *Science*. 2006;313:504–507.

[9] LeCun Y, Bengio Y, Hinton G. Deep learning. *Nature*. 2015;521:436–444.

[10] McCulloch WS, Pitts W. A logical calculus of the ideas immanent in nervous activity. *Bull Math Biol*. 1990;52:99–115.

[11] Cortes C, Vapnik V. Support-vector networks. Mach Learn. 1995;20:273–297.

[12] Einerson CJ, Clark DE, Detering B. Intelligent control strategies for the plasma spray process. *Therm Spray Coat Res, Des Appl, Proc Natl Spray Conf.* 1993. p. 205–211.

[13] Singh J, Singh S. A review on machine learning aspect in physics and mechanics of glasses. *Mater Sci Eng B* [Internet]. 2022;284:115858. Available from: https://doi.org/10.1016/j.mseb.2022.115858.

[14] Praveen AS, Sarangan J, Suresh S, et al. Optimization and erosion wear response of NiCrSiB/WC-Co HVOF coating using Taguchi method. *Ceram Int.* 2016;42:1094–1104.

[15] Qiao L, Wu Y, Hong S, et al. Relationships between spray parameters, microstructures and ultrasonic cavitation erosion behavior of HVOF sprayed Fe-based amorphous/nanocrystalline coatings. *Ultrason Sonochem.* 2017;39:39–46.

[16] Singh J, Kumar S, Singh G. Taguchi's approach for optimization of tribo-resistance parameters Forss304. *Mater Today Proc* [Internet]. 2018;5:5031–5038. Available from: https://doi.org/10.1016/j.matpr.2017.12.081.

[17] Tabbara H, Gu S, McCartney DG. Computational modelling of titanium particles in warm spray. *Comput Fluids.* 2011;44:358–368.

[18] Li M, Christofides PD. Modeling and control of high-velocity oxygen-fuel (HVOF) thermal spray: A tutorial review. *J Therm Spray Technol.* 2009;18:753–768.

[19] Dongmo E, Wenzelburger M, Gadow R. Analysis and optimization of the HVOF process by combined experimental and numerical approaches. *Surf Coat Technol.* 2008;202:4470–4478.

[20] Guessasma S, Salhi Z, Montavon G, et al. Artificial intelligence implementation in the APS process diagnostic. *Mater Sci Eng B.* 2004;110:285–295.

[21] Guessasma S, Montavon G, Coddet C. Modeling of the APS plasma spray process using artificial neural networks: Basis, requirements and an example. *Comput Mater Sci.* 2004;29:315–333.

[22] Guessasma S, Montavon G, Coddet C. Neural computation to predict in-flight particle characteristic dependences from processing parameters in the APS process. *J Therm Spray Technol.* 2004;13:570–585.

[23] Sahraoui T, Guessasma S, Fenineche NE, et al. Friction and wear behaviour prediction of HVOF coatings and electroplated hard chromium using neural computation. *Mater Lett.* 2004;58:654–660.

[24] Choudhury TA, Hosseinzadeh N, Berndt CC. Artificial Neural Network application for predicting in-flight particle characteristics of an atmospheric plasma spray process. *Surf Coat Technol.* 2011;205:4886–4895.

[25] Choudhury TA, Hosseinzadeh N, Berndt CC. Improving the generalization ability of an artificial neural network in predicting in-flight particle characteristics of an atmospheric plasma spray process. *J Therm spray Technol.* 2012;21:935–949.

[26] Kanta A-F, Montavon G, Planche M-P, et al. Artificial neural networks implementation in plasma spray process: Prediction of power parameters and in-flight particle characteristics vs. desired coating structural attributes. *Surf Coat Technol.* 2009;203:3361–3369.

[27] Kanta A-F, Montavon G, Berndt CC, et al. Intelligent system for prediction and control: Application in plasma spray process. *Expert Syst Appl.* 2011;38:260–271.

[28] Kanta A-F, Montavon G, Planche M-P, et al. Artificial intelligence computation to establish relationships between APS process parameters and alumina-titania coating properties. *Plasma Chem Plasma Process.* 2008;28:249–262.

[29] Zhang C, Kanta A-F, Li C-X, et al. Effect of in-flight particle characteristics on the coating properties of atmospheric plasma-sprayed 8 mol% Y2O3-ZrO2 electrolyte coating studying by artificial neural networks. *Surf Coat Technol.* 2009;204:463–469.

[30] Liu T, Deng S, Planche M-P, et al. Estimating the behavior of particles sprayed by a single-cathode plasma torch based on a nonlinear autoregressive exogenous model. *Surf Coat Technol.* 2015;268:284–292.

[31] Guessasma S, Montavon G, Gougeon P, et al. Designing expert system using neural computation in view of the control of plasma spray processes. *Mater Des*. 2003;24:497–502.

[32] Kanta A-F, Planche M-P, Montavon G, et al. In-flight and upon impact particle characteristics modelling in plasma spray process. *Surf Coat Technol*. 2010;204:1542–1548.

[33] Cherigui M, Guessasma S, Fenineche N, et al. Neural computation to correlate HVOF thermal spraying parameters with the magnetic properties of FeNb alloy deposits. *Mater Chem Phys*. 2005;93:181–186.

[34] Zhang G, Kanta A-F, Li W-Y, et al. Characterizations of AMT-200 HVOF NiCrAlY coatings. *Mater Des*. 2009;30:622–627.

[35] Kamnis S, Malamousi K, Marrs A, et al. Aeroacoustics and artificial neural network modeling of airborne acoustic emissions during high kinetic energy thermal spraying. *J Therm Spray Technol*. 2019;28:946–962.

[36] Mojena MAR, Roca AS, Zamora RS, et al. Neural network analysis for erosive wear of hard coatings deposited by thermal spray: Influence of microstructure and mechanical properties. *Wear*. 2017;376:557–565.

[37] Liu M, Yu Z, Zhang Y, et al. Prediction and analysis of high velocity oxy fuel (HVOF) sprayed coating using artificial neural network. *Surf Coat Technol*. 2019;378:124988.

[38] Liu M, Yu Z, Wu H, et al. Implementation of artificial neural networks for forecasting the HVOF spray process and HVOF sprayed coatings. *J Therm Spray Technol*. 2021;30:1329–1343.

[39] Liu M, Wu H, Yu Z, et al. Description and prediction of multi-layer profile in cold spray using artificial neural networks. *J Therm Spray Technol*. 2021;30:1453–1463.

[40] Wang Z, Cai S, Chen W, et al. Analysis of critical velocity of cold spray based on machine learning method with feature selection. *J Therm Spray Technol*. 2021;30:1213–1225.

# 2 Unsupervised and Supervised Machine Learning Techniques in Wear Prediction

## ABBREVIATIONS

| | |
|---|---|
| **AI** | Artificial intelligence |
| **ANN** | Artificial neural networks |
| **CNN** | Convolutional neural network |
| **HVOF** | High-velocity oxy-fuel |
| **I/P** | Input |
| **ML** | Machine learning |
| **O/P** | Output |
| **PS** | Plasma spray |
| **TS** | Thermal spraying |
| **SML** | Supervised machine learning |
| **USML** | Unsupervised machine learning |

## 2.1 INTRODUCTION TO MACHINE LEARNING IN WEAR ANALYSIS

Metals are subjected to wear through the plastic deformation of the material and particle detachment in the form of wear debris. Wear may be of a mechanical type such as adhesion, abrasion, and erosion. It can also be a chemical type, which is commonly known as corrosion. Wear is the loss of material from a solid surface caused by its interaction with another solid surface [1]. The different wear issues that affect different industries are represented in Table 2.1. The theory of tribology includes the study of wear. Being present in practically every part of our lives, tribology has been and continues to be one of the most pertinent topics in today's society. The considerable portion of the world's energy consumption in the present day [2] also illustrates the significance of friction, lubrication, and wear. Wear-related problems can be solved creatively with the use of tribology knowledge. Numerous precise tests and cutting-edge computer simulations conducted at various scales and in a variety of physical disciplines form the foundation of all advancements [3]. Advanced data handling, evaluation, and learning models can be created based on this strong and data-rich basis and used to extend current knowledge in the context of tribology 4.0 [4] or tribo-informatics [5]. Additionally, tribology is defined by the fact that the underlying

DOI: 10.1201/9781003400660-2

**TABLE 2.1**
**Different Wear Problems in Industries [1]**

| S. No. | Wear Problems Identify in Industries | Significant Characteristics | Examples |
|---|---|---|---|
| 1 | Surface wear caused by hard particles in a fluid stream | Erosion | Flow-controlling valves for crude oil |
| 2 | Surface wear by abrasive particles in a compliant material bed | Abrasion, with the abrasive supply being continuously refreshed by the movement of the material bed | Teeth of diggers, extrusion dies for tiles |
| 3 | Wear of metal surfaces when they rub against one another in the presence of abrasive particles | Three body abrasion | Pivot pins used in construction machinery, seals for shaft containing abrasive particles |
| 4 | The wear of components made from metal via rubbing contact with a sequence of other solid components | Adhesive and abrasion wear, although one part of the wear process is continuously renewed | Sintering dies, cutter blades, punching tools |
| 5 | The wear of metal component pairs in mutual and repeated contact | Adhesive wear | Piston rings and liners for cylinders |
| 6 | Component wear is caused by metals and nonmetals rubbing against each other | Adhesive wear | Brakes and clutches, artificial hip joints |

*Source:* Permissions used Creative Commons Attribution 3.0 License.

processes/science behind tribology cannot yet be fully described mathematically, such as by differential equations. Therefore, artificial intelligence (AI) coupled with well-known machine learning (ML) techniques from the computer science field is broadly affecting many aspects of Industrial Revolution 4.0, including science and technology, the manufacturing industry, and even our day-to-day lives. The ML methods have been designed to analyze a large amount of data to gain insightful information, classify, predict, and make judgments based on evidence in unique ways. Traditional domain-specific modeling and simulation skills have been accelerated, improved, and completed with the use of ML algorithms in the last decade. The benefits and promise of ML and AI techniques may be observed, in particular, in their capacity to handle high-dimensional problems with a reasonable amount of work and expense [6].

ML was first considered a branch of computer science but is now a component of AI [7]. Logic, algorithm theory, probability theory, and computing form AI and ML [8]. Designing computing systems for a particular activity that can gradually learn from training data and create and improve knowledge-based models that predict outcomes is the initial step in ML. Thus, questions in the relevant field can be answered using ML [7]. This will encourage the development of novel algorithms that support AI's sustained growth in modern-day industries. A variety of different algorithms can be applied to ML, with their applicability being significantly task-dependent.

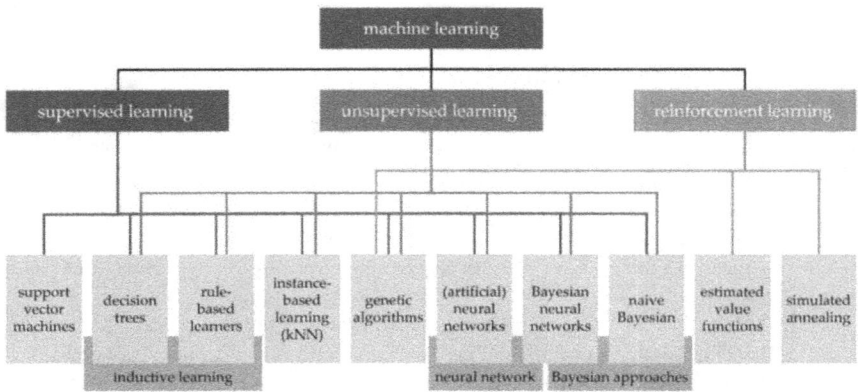

**FIGURE 2.1**  Classification of various ML techniques [6]. Permissions under the Creative Commons Attribution 4.0 License.

Algorithms can generally be divided into two categories: "supervised machine learning" (SML) and "unsupervised machine learning" (USML) [8]. Classification of various ML techniques is represented in Figure 2.1. This chapter summarizes the various models, mainly artificial neural network (ANN), and convolutional neural network (CNN) under SML and USML approaches which were used for the prediction of wear in components/parts. These ML models offer a strong and practical method for modeling the wear of industrial components.

## 2.2  NEED FOR MODELING TECHNIQUES TO PREDICT WEAR RATE OF THERMAL SPRAY COATINGS

There are numerous complex physical phenomena connected to the coating procedure in thermal spraying (TS). Complex nonlinear interdependencies between process parameters, in-flight particle characteristics, and coating structure are present during TS procedures. To quantify these complicated relationships and improve process repeatability, computer-aided approaches are used. To comprehend these interactions, conventional modeling techniques are frequently used. This necessity has inspired researchers to create a reliable methodology that uses modern digital techniques like AI to resolve the problems of TS-deposited coatings under various operating conditions. The purpose is to deeply discuss AI, and various models related to AI, such as ANN, CNN, and hyperspectral imaging, that can promote the digital transformation of the modern surface engineering industry and help in achieving a durable TS coating system. To encourage the use of such reliable ML approaches for process modeling and TS parameter optimization, a summary of the roles of ANNs and CNNs has been discussed in this chapter.

## 2.3  ANN TECHNIQUE IMPLEMENTED IN WEAR PREDICTION

The networks of biological neurons that form human brains inspire ANN computer models. The dendrites serve as the input (I/P) vector, as shown in Figure 2.2,

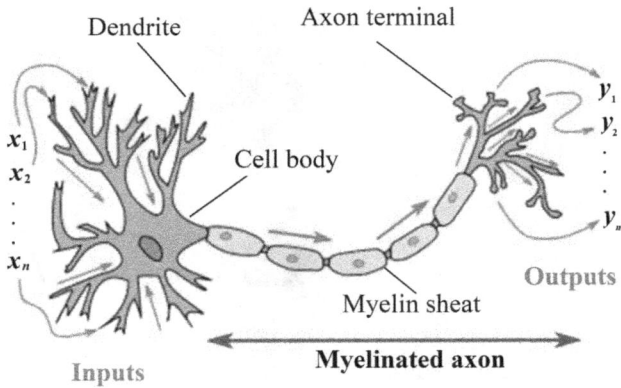

**FIGURE 2.2** Signal flow between input at dendrites and output at axon terminals [10]. Permissions under the Creative Commons Attribution 4.0 license.

allowing the cell body (or soma) to receive signals from a significant number of nearby neurons. From the neuron, axons transmit signals to neighboring cells. The cell body processes the dendrites' signals, which serves as a summing function. The neuron pumps sodium or potassium in and out depending on how the environment and cell body interact, changing the electrical potential of the neuron. A neuron "fires," producing an action potential that travels down the axons to the synapses and other neurons after its electrical potential reaches a specific potential [9].

Similar to a biological neuron, ANN works by mapping an I/P space to an output (O/P) space using a set of operating elements called neurons and connections known as weights (analogous to synapses in a biological brain). A transfer function that mirrors the action potential's firing rate describes how neurons transmit signals to other neurons by sending action potentials down their axons. Different I/Ps to the neuron may be more or less relevant, depending on factors like whether the neuron should fire, causing them to have a smaller or bigger impact. It is accomplished by changing the weight. As a result, the neuron can be thought of as a little computer that receives I/P, processes them, and sends out an O/P. An elementary neuron with 'R' I/Ps ($p_1$ $p_2$.....$p_R$) a corresponding transfer function $f$, and an O/P $a$ are shown in Figure 2.3. Each I/P is given the proper weight ($w$). The bias and the weighted I/P added together make up the transfer function's I/P.

To find a relationship between I/P and O/P, learning algorithms are needed. Several learning algorithms are accessible for processing; however, the feed-forward-back propagation technique is the most popular approach. In the back-propagation learning algorithm, the relationship between I/P and O/P is determined by a nonlinear transfer function.

The most popular nonlinear transfer function when creating an ANN network is the sigmoid function, which is represented in the following equation [10]:

$$F(x) = \frac{1}{1 + e^{-x}} \tag{2.1}$$

**General Neuron**

**Input**

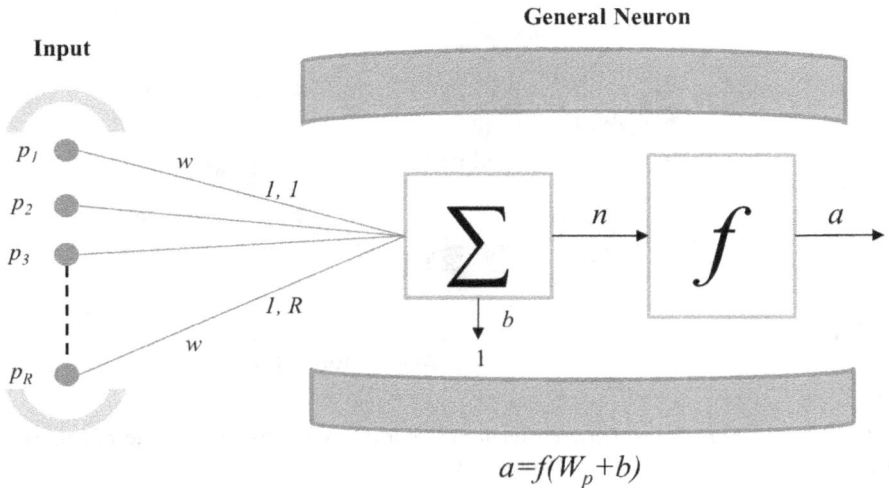

$$a = f(W_p + b)$$

**FIGURE 2.3**  An elementary artificial neuron's structure with $R$ inputs.

Equation 2.1 defines log sigmoid functions. The log sigmoid and hyperbolic tangent sigmoid transfer (sigmoidtransfer) functions are two variants of nonlinear sigmoid transfer functions used to determine the relationship between I/P and O/P (tansig). Equation 2.2 represents the Tansig function [11]:

$$F(x) = \frac{1 - e^{-2x}}{1 + e^{-2x}} \qquad (2.2)$$

The model is trained using variables such as momentum factor, rate of learning, hidden layers numbers, and the number of neurons in the hidden layers. The root mean square error, i.e., RMSE, between the O/P and expected values is determined for performance evaluation. The ANN estimates the error by comparing the predicted values to the target values after each iteration. If the error exceeds the allowed error, the network is run again while the weights are changed to decrease the error. The data is separated into test, training, and validation data sets to prevent the overfitting issue. Figure 2.4 displays the streamlined ANN procedure for generating the desired result.

## 2.4  NEURAL COMPUTATION OR BASIC ANN STRUCTURE FOR TS COATINGS

An intelligent system that links the processing parameters to the process responses (i.e., characteristics of the in-flight particles) is introduced as neural computation. Such a system is built on an ANN, which is a network of interconnected neurons that are processing units. ANNs, which are fast-evolving technologies with flexible topologies and strong learning capabilities, have been used in various engineering fields [12–14]. ANN is essentially split into single-layer neural networks and multilayer networks. Due to its excellent accuracy, the multilayered feed-forward network is the most commonly used ANN model for TS applications. Few articles published

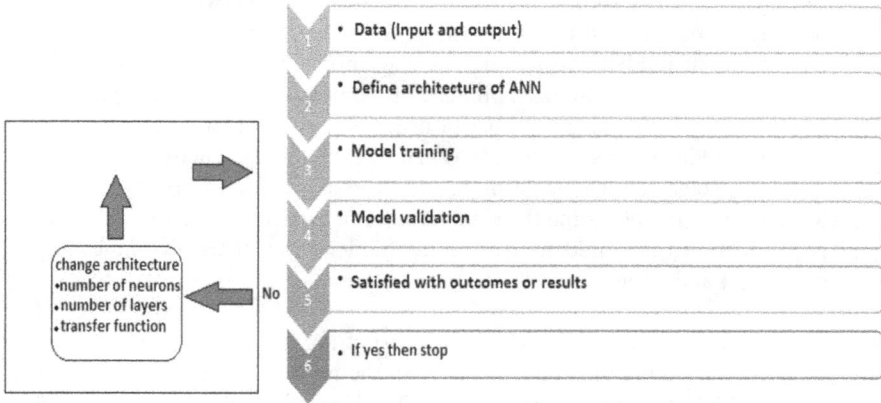

• Data (Input and output)

• Define architecture of ANN

• Model training

• Model validation

change architecture
•number of neurons
•number of layers
•transfer function

No

• Satisfied with outcomes or results

• If yes then stop

**FIGURE 2.4**   Streamlined artificial neural network procedure.

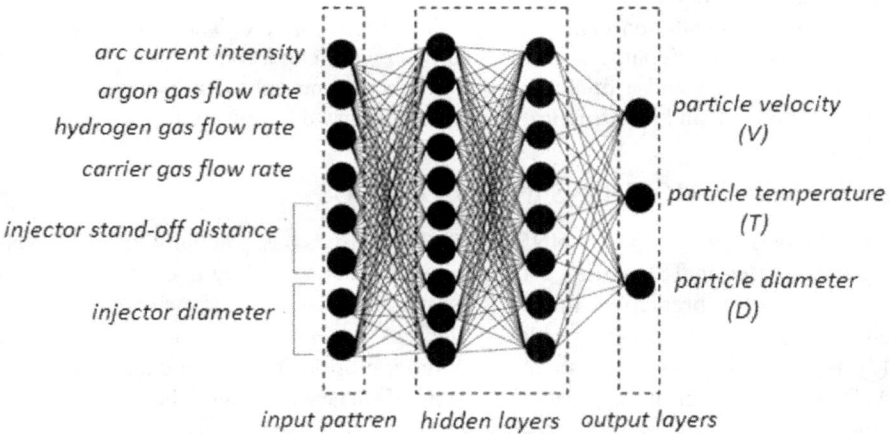

arc current intensity
argon gas flow rate
hydrogen gas flow rate
carrier gas flow rate

injector stand-off distance

injector diameter

particle velocity
(V)

particle temperature
(T)

particle diameter
(D)

input pattren   hidden layers   output layers

**FIGURE 2.5**   A typical layout representing a multilayer artificial neural network structure [17]. Permissions from Elsevier.

on surface coating applications utilized a multilayered model [15]. Typically, a multilayered neural network has three layers, as mentioned below:

• The pattern of I/P represents the settings for process parameters. Depending on the type of parameter, one or more neurons describe each parameter in this way. One neuron is needed to describe a single actual value, such as the flow rate of argon gas, while $x$ neurons are needed to explain $2^x$ categories connected to the parameter, such as injector type. The size of the I/P pattern may vary depending on the I/P parameters selected. In Guessasma et al.'s [16] work, the I/P pattern size is 8 (inset Figure 2.5).
• The O/P pattern shows the temperature, diameter, and particle velocity. Each neuron describes one of these parameters.

- The hidden layers subtly describe the relationships between the in-flight particle properties and the processing parameters. The precise number of neurons in the hidden layers cannot be determined by any generic rules. The flexibility of having a higher number of parameters to optimize is provided by a network with a large number of neurons in the hidden layer. However, once the hidden layer size exceeds a specific threshold, the network becomes undercharacterized. Underfitting results from the hidden layers having too few neurons. To determine the ideal number of hidden layers, the simulation is run with just one hidden layer. In the hidden layer, there are somewhere between 4 and 20 neurons.

A number termed weight that translates the strength of the connection between two neurons is used to define the connection. With the help of a transfer function, each neuron in the ANN structure transforms the flux it receives from the other neurons, which is a weighted number. This transformation makes sure that the process is non-linear at the scale of each neuron. The learning process used for the ANN structure optimization takes into account experimental sets. The number of neurons and the weight of the population can be fixed using this technique. The system response is computed and compared with the findings of the experiment using an assumed weight population and a predetermined number of neurons. Each layer's residual error is assessed, and the structure weights are adjusted to reduce it.

## 2.5   ANN APPLICATIONS IN TS COATINGS

Over the past several years, ANN's role in TS coatings has gradually grown. ANNs are widely used in TS coatings for a variety of tasks, including microhardness prediction, porosity prediction, thickness estimation, coating microstructure analysis, prediction of tribological properties of coatings, and in-flight particle behavior [16, 17]. By excluding undesired sounds and making up for the manipulated variables, ANN is a key player in the process modeling of surface coatings. These were able to more precisely optimize such intricate nonlinear processes and forecast more favorable circumstances for coating deposition on the material's surface. For the selection of TS process parameters in the future, notably for the design of ceramic coatings with specific functional qualities, the merging of AI methods is crucial. Recently, ANN has also been used for modeling track profiles during additive manufacturing via cold spray. Future research will include the data-efficient ANN model in the toolpath planning algorithm to enhance geometric control and produce more complicated product designs via cold spray additive manufacturing [18,19]. Table 2.2 summarizes the various applications of ANN in TS coatings, and Figure 2.6 shows the distribution of articles according to how they are used in surface coating techniques.

## 2.6   CONVOLUTIONAL NEURAL NETWORK

CNN is a particular kind of ANN frequently used for visual images. These networks perform well with data that has a grid-like pattern. As the name suggests, CNN uses the convolutional mathematical procedure. CNN can automatically mine potential

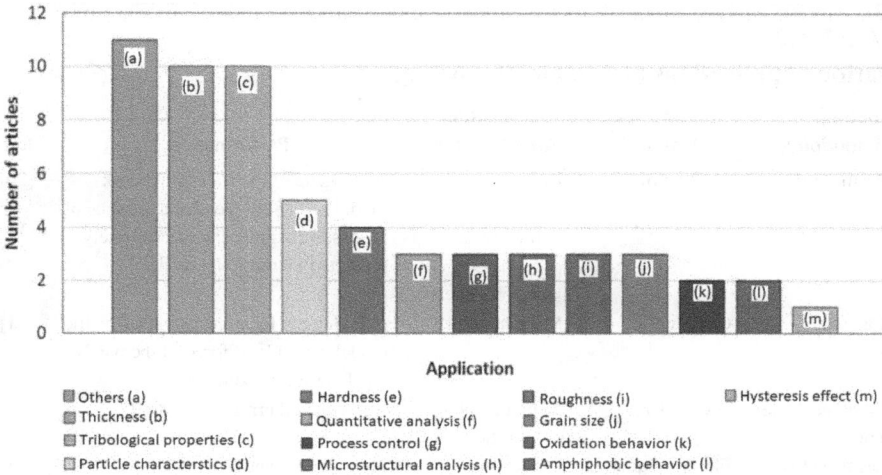

**FIGURE 2.6**  Article distribution based on application areas *y* based on ML [15]. Permissions from Elsevier.

pattern characteristics. It exhibits remarkable performance and has been successfully used in image-processing fields like target identification and face recognition. CNN uses several filters, and each filter collects data from the image, like edges and various forms (vertical, horizontal, and round), before combining all of this information to identify the image. It is not possible to achieve similar results using ANNs. This is due to various demerits associated with ANN:

- The amount of computing required to train an ANN model on large images and various image channels is too much;
- The second drawback is that, in contrast to a CNN model, it is unable to collect all of the information from an image, including its spatial dependencies;
- Another problem is that ANN is sensitive to the object's placement in the image; as a result, it will be unable to correctly classify an object if its location changes.

CNN has recently received attention in the realm of materials engineering. For instance, the CNN model has been utilized successfully to predict the relationships between material microstructure and property [26–28]. To investigate the connection between processing conditions and microstructure and to comprehend the high-dimensional microstructure representation, the VGG16 model, which was trained on ImageNet, was used [29]. Deep learning can be used to operate on the segmentation of material images [30], classification of microstructure images [31–33], and reconstruction of material microstructures [34,35]. The CNN model exhibits strong performance in feature extraction thanks to its impressive learning capacity. To learn the potential association between particle distributions and control parameters in the plasma spray process, the parameters in filters and fully connected layers were trained [36–39].

**TABLE 2.2**

**Various Applications of ANN in TS Coatings**

| Coating Methodology | Feedstock Material | Application Area Targeted | Performance | Ref. |
|---|---|---|---|---|
| Plasma spray | Ni60CuMo | Bonding strength, porosity, and coating microhardness | The $R$-value of the trained network model is 0.8828; the characteristics of Ni-based coatings can be accurately predicted by the ANN model | [20] |
| Cold spray | Copper | Microhardness | Following the local thermal history, the model was able to forecast the part's local hardness reduction | [21] |
| HVOF & flame spray | WC-CoCr | Tribological properties | Mean squared error – 0.000689 | [22] |
| Plasma spray | Flay-ash, quartz, and ilmenite | Deposition efficiency | The outcomes show that neural network analysis can produce results that are quite accurate and can be utilized as a useful tool in the production process for plasma deposition | [23] |
| Plasma spray | CoMoCrSi | Coating porosity | The porosity drops to just 5.6% after the ANN/genetic algorithm optimization process | [24] |
| Plasma spray | $Al_2O_3 + 13\%TiO_2$ | Hardness, porosity, and cavitation erosion resistance | An innovative strategy is the combination of genetic algorithms with ANN. They discovered a set of Pareto-optimal solutions by using multiobjective optimization | [25] |
| Cold spray additive manufacturing | Titanium | Modeling of track profile | The findings show that a neural network model can perform better than a popular mathematical model using data-efficient modeling techniques and be more suitable for enhancing geometric control in CSAM | [18] |
| Cold spray additive manufacturing | | Geometric modeling | The findings show that when combined with the right process planning algorithm, a neural network modeling approach is well suited for predicting cold spray profiles and may be utilized to enhance geometric control in AM | [19] |

ANN, artificial neural network, TS, thermal spraying

### 2.6.1 CONVOLUTION

The definition of the convolution operation is given by Equation 2.3:

$$s(t) = (x * w)(t) = \int_{-\infty}^{\infty} x(\tau) w(t - \tau) d\tau \tag{2.3}$$

where $w$ is frequently referred to as the feature map or kernel, and function $x$ is frequently used as the I/P. We want our kernel $K$ to be 2D if the I/P picture $I$ is 2D. Consequently, we can characterize our convolutional operations as:

$$S(i,j) = (K \times I)(i,j) = \sum_{m} \sum_{n} I(m,n) K(i - m, j - n) \tag{2.4}$$

where $m$ and $n$ have varied overall positions in the kernel $K$.

### 2.6.2 CONSTRUCTION OF THE CNN MODEL

Feature extraction and classification are the two main phases of a CNN model's operation. In the feature extraction step, different filters and layers are used to extract information and features from the images. After this step is finished, the images move onto the classification phase, where they are labeled based on the target variable of the underlying problem (in Figure 2.7).

For thermal spray coatings, it is first important to recognize and recover the image features associated with the coating's service performance to accurately construct a quantitative relationship model. CNN recognizes these characteristics, which have the following layers: I/P, convolution, pooling, fully connected, and O/P.

- *I/P layer:* The I/P image, as the name sounds, can be either Red Green Blue (RGB) or Grayscale schemes. Each image is composed of pixels. Before sending them to the model, we must normalize them in the range from 0 to 1.

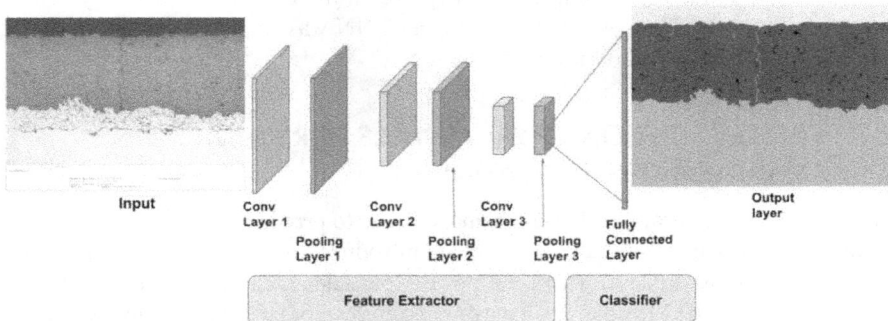

**FIGURE 2.7** Generalized CNN model [36]. Permissions under the Creative Commons Attribution 4.0 license.

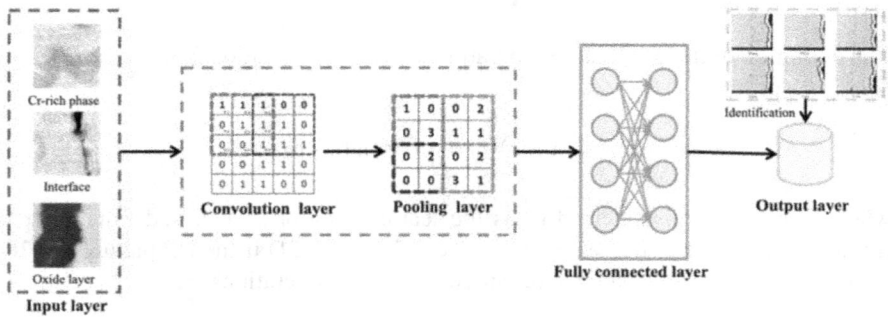

**FIGURE 2.8**  Typical phases of the NiCrAlY coating following oxidation are recognized by CNN's architecture [36]. Permissions under the Creative Commons Attribution 4.0 license.

- *Convolution layer:* The filter is applied to the I/P image during the convolution layer to extract or detect its features. Multiple filtering operations are done on the image.
- *Pooling layer:* Following the convolutional layer, the pooling layer is used to decrease the feature map's size, aiding in the preservation of the I/P image's key details or features while speeding up computation.

Figure 2.8 represents the outline of the CNN model to recognize the NiCrAlY coating characteristics phases after oxidation testing. Among these, the I/P layer is an RGB-channeled, $64 \times 64$-pixel image. To train the I/P image, an 8-CNN network with 3 convolutional layers and 32 convolution kernels in every convolutional layer is chosen [36]. A pooling layer comes after each convolutional layer. A $2 \times 2$ sliding filter is used in conjunction with the maximum pooling method to increase data processing effectiveness. Regularized Dropout random neuron inactivation and the SoftMax classifier are employed in the fully connected layer to improve the network's anti-interference capabilities and lessen over-fitting. Binary image processing and flood-filling technologies were utilized to choose the right threshold to extract the developed oxide layer from the coating's scanning electron microscopy (SEM) images, and its thickness was calculated. The CNN was implemented using Keras, an open-source Python toolkit for ANNs.

## 2.7   IMPLEMENTATION OF ML MODELS FOR PREDICTION OF WEAR RATES IN COATINGS

The main purpose of atmospheric plasma spray is to provide protective coatings for metallic and ceramic materials. Powders are introduced into a hot plasma plume during this deposition process, where the powder particles melt and are then deposited onto a substrate. The Mo coatings have drawn interest for applications involving high temperatures [40]. Because of their remarkable wear resistance characteristics, PS Mo-based coatings are mostly selected for the aerospace, pulp, automotive, and paper industries [41]. Cetinel et al. [42] conducted wear tests on Mo coatings deposited by plasma spraying using a universal wear-friction setup. The wear behavior of PS Mo coatings has been simulated using a neural network technique (ANN). The I/P layer, two hidden layers, and the O/P layer make up the neural network in this study. The dry and acidic atmosphere, normal load, and timing were the I/P variables. Microhardness and wear loss were considered the two O/P variables. The Levenberg-Marquardt approach was used to train and implement the ANN using MATLAB neural network functions [43].

The authors concluded that only a few milliseconds were required for calculations performed during the neural network testing phase. As a result, ANN may be used to analyze the Mo coatings' wear process and calculate the amount of wear loss without consuming time. Satpathy et al. [44] also report the successful implementation of ANN to examine the wear behavior of plasma spray coatings. The authors recommended that ANN is a suitable method to save time and resources. Pati et al. [45] combine the use of experimental design and ANN to examine the erosion and wear behavior of plasma-sprayed coatings. The authors found a good agreement was established between the experimental and ANN-predicted data, keeping the error to within 7%. Table 2.3 summarizes different studies on TS coatings using ML models.

**TABLE 2.3**
**Different Studies on TS Coatings Using ML Models**

| TS Method | Feedstock Composition | I/P Parameters | O/P Parameters | Ref. |
|---|---|---|---|---|
| Plasma spray | $Al_2O_3$ + 13%$TiO_2$ | • Stand-off distance<br>• Torch velocity | • Microhardness<br>• Porosity<br>• Cavitation erosion resistance | [25] |
| Cold spray | Cu, Al, Al6061, Al7075-T6, Ti, Ni, Fe & TC4 | • Tensile strength<br>• Yield strength<br>• Thermal conductivity<br>• Sound speed<br>• Melting point | • Critical velocity | [46] |
| Plasma spray | WC-12Co | • The intensity of arc current, a flow rate of $H_2$, and Ar flow rate for model 1<br>• The intensity of arc current, a flow rate of $H_2$, Ar flow rate, the temperature of particles, and velocity of particles for Model 2 | • The temperature of particles, and velocity of particles for for model 1<br>• Porosity and the hardness of coating for model 2 | [47] |
| Plasma spray | Zirconia | • Spraying layers, travel speed, voltage, powder feed rate, Arc current, stand-off distance, the flow rate of carrier gas, and the flow rate of primary gas | • Wear loss volume | [48] |

*(Continued)*

**TABLE 2.3 (*Continued*)**

**Different Studies on TS Coatings Using ML Models**

| TS Method | Feedstock Composition | I/P Parameters | O/P Parameters | Ref. |
|---|---|---|---|---|
| Plasma spray | Ni-5 wt%Al | • Arc current, stand-off distance, powder feed rate, and the flow rate of primary gas | • Microhardness, thickness & coating porosity | [49] |
| Plasma spray | CoMoCrSi | • Spray distance, the pressure of the chamber, the current, flow rate of Ar gas, and flow rate of $H_2$ gas | • Coating porosity | [24] |
| Plasma spray | YSZ | • I/P power, primary gas flow rate, stand-off distance, powder feed rate, and the carrier gas flow rate | • Deposition efficiency, tensile bond strength, hardnss, and surface roughness | [50] |
| Plasma spray | $Al_2O_3 + 13\%TiO_2$ | • The flow rate of carrier gas<br>• Diameter of the injector<br>• Injector stand-off distance<br>• $H_2$ flow rate<br>• Ar flow rate<br>• The intensity of the arc current | • The temperature of particles<br>• The velocity of particles<br>• Diameter of particles | [17,51] |
| Plasma spray | $Al_2O_3 + 13\%TiO_2$ | • The intensity of the arc current<br>• $H_2 + Ar$<br>• $H_2/Ar$<br>• Diameter of the injector<br>• The flow rate of carrier gas | • Height & flattening of the deposited coating profile | [52] |
| Plasma spray | $Al_2O_3 + 13\%TiO_2$ | • The intensity of arc current<br>• $H_2 + Ar$<br>• $H_2/Ar$<br>• The flow rate of carrier gas | • The temperature of particles<br>• Velocity of particles | [53] |
| Plasma spray | $Al_2O_3 + 13\%TiO_2$ | • Temperature of particles<br>• Velocity of particles | • Porosity | [54] |
| Plasma spray | Mo powders | • Wear conditions (acidic and dry)<br>• Normal load<br>• Wear time | • Wear loss & microhardness | [55] |
| HVOF & flame spray | WC-Co-Cr $Cr_3C_2$-NiCr | • Method of deposition<br>• Impingement angle<br>• Abrasive particles velocity<br>• Porosity<br>• Fracture toughness<br>• Microhardness<br>• Surface roughness<br>• Density of coatings | • Erosion rate | [22] |
| HVOF | NiCrAlY | • Oxygen flow rate<br>• Stand-off distance | Coating microhardness & porosity | [56] |
| HVOF | $Cr_3C_2$-NiCr | • Oxygen flow rate<br>• Stand-off distance<br>• $CH_4$ flow rate | Coating microhardness, wear rate & porosity | [57] |

## 2.8    AI IN WEAR MONITORING OF WIND TURBINE BLADES

The wind energy industry as a whole faces the issue of material erosion in wind turbine blades as a result of the effects of weather. Using AI, scientists were able to devise a workable answer to the material erosion issue. Material erosion on wind turbine blades is an unexpectedly expensive problem that has so far defied solution. Wind turbines have a short lifespan because the blade material deteriorates when exposed to bad weather. Because of this issue, the value of all wind-generated power can be reduced by as much as 2%–4%. It's a timeless issue that plagues this sector of the economy, increasing prices for all forms of wind power by billions of Euros [58]. Many techniques have been used in recent years to forecast the functionality and state of wind turbine monitoring systems. ANNs are being used to assess performance in real-time, which can be tapped into for the purpose of enhancing fault detection systems [59]. Many researchers have employed ANNs and CNNs to describe the typical behavior of wind turbines. Yang and Cheng [60] used CNNs and ANNs in the suggested methodology. While ANNs draw attention to surface damage, CNNs categorize images of surface damage as either positive or negative. Whereas the ANN is highly trained with its data using feature-based training, CNN is well trained with its data using image-based training. An ANN model has a 70.7% accuracy compared to a CNN model's 89.4% accuracy.

## 2.9    IMAGE PROCESSING TECHNIQUES

Drones, according to Moreno-Armendári et al. [61], would improve the effectiveness of wind turbine blade maintenance. With a camera built inside the drone, CNN can recognize and categorize different sorts of damage in photos captured by the gadget. In this work, the accuracy metric was used by the authors to evaluate the effectiveness of a system for detecting blade damage. Mavic 2 Pro UAVs were used by Xu et al. [62] to take pictures of the wind turbine blades. A total of 25,773 photos at a resolution of $5472 \times 3684$ of the wind turbine blade have been acquired from UAVs. With the aid of their UAV photos, Wang and Zhang [63] presented their groundbreaking investigation into spotting blade flaws in the wind turbine. Using photos of wind turbine blades, data-driven algorithms are created to evaluate and pinpoint the characteristics of blade defects. Hence, many applications for predictive maintenance, both in wind turbines and elsewhere, can be made use of these methodologies.

## 2.10    HYPERSPECTRAL IMAGING (HSI) FOR WEAR
## DETECTION OF BLADES

Although image processing methods eliminate the need for human contact with the blades for fault detection, imaging is still done remotely and does not always produce high-quality images using typical HD cameras. As a result, imaging methods based on other parts of the electromagnetic spectrum are becoming more popular. To fully evaluate this potential and comprehend which frequencies can be used with imaging blades, laboratory and field studies are necessary. A portion of a wind turbine blade with surface flaws is imaged using the HSI technique [64]. This method for remote in-field inspections offers great accuracy in a shortened inspection period. Surface and subsurface imperfections may be quantified and localized at an early stage of formation [65].

## 2.11   CONCLUSION AND FUTURE PERSPECTIVE

The authors have thoroughly analyzed the digital technologies available for modeling to achieve high-performance coatings. AI-related models like ANN and CNN are given more attention. The key outcomes are as follows:

- It is anticipated that with suitable models and ML technologies, it will be possible to create coatings with certain target properties using fewer tests.
- An I/P–O/P relationship is mapped using an ANN, a feed-forward network model for SML, depending on relevant training data.
- Several pre-processing aspects, such as choosing relevant I/P variables, data quality, and network architecture, affect the formation of an ANN with acceptable prediction accuracy.
- CNNs can also be used to classify real-time video and images of spray processes, where less calibration is needed due to the presence of enormous datasets. The approach is simple to implement, but larger research efforts are required to produce huge datasets and kernels.
- ANN is also used for modeling track profiles during additive manufacturing via cold spray. Future research will include the data-efficient ANN model in the tool-path planning algorithm to enhance geometric control and produce more complicated product designs via cold spray additive manufacturing.
- Further research utilizing AI methods based on deep learning or ML algorithms has to be done.
- The next method of inspecting wind turbine blades, known as HSI, will result in a shorter inspection shutdown time, lower maintenance costs, and a lower frequency of unexpected failures by offering a simple, routine inspection of the blade.
- In terms of future work, HSI will be put to the test on various blade material types to research the impact of manufacturing material on the detection process.

## REFERENCES

1. Swain B, Bhuyan S, Behera R, Mohapatra SS, Behera A. 2020. Wear: A serious problem in industry. In: Patnaik A, Singh R, Kukshal V editors. *Tribology in Materials and Manufacturing-Wear, Friction and Lubrication.* London: IntechOpen.
2. Holmberg K, Erdemir A. Influence of tribology on global energy consumption, costs and emissions. *Friction.* 2017;5:263–284.
3. Vakis AI, Yastrebov VA, Scheibert J, Nicola L, Dini D, Minfray C, Almqvist A, Paggi M, Lee S, Limbert G, Molinari JF. Modeling and simulation in tribology across scales: An overview. *Tribol Int.* 2018;125:169–199.
4. Ciulli E. Tribology and industry: From the origins to 4.0. *Front Mech Eng.* 2019;5:55.
5. Zhang Z, Yin N, Chen S, Liu C. Tribo-informatics: Concept, architecture, and case study. *Friction.* 2021;9:642–655.
6. Wuest T, Weimer D, Irgens C, Thoben KD. Machine learning in manufacturing: Advantages, challenges, and applications. *Prod Manuf Res.* 2016;4(1):23–45.
7. Bell J. *Machine Learning: Hands-On for Developers and Technical Professionals.* Hoboken, NJ: Wiley; 2014. ISBN 978-1-118-88906-0.
8. Wittpahl V. *KünstlicheIntelligenz.* Berlin, Germany: Springer; 2019. ISBN 978-3-662-58041-7.

9. Priddy KL, Keller PE. *Artificial Neural Networks: An Introduction*, Vol. 68. SPIE Press; 2005.

10. Vu-Quoc ESPL. https://commons.wikimedia.org/wiki/File:Neuron3.png; 2018.

11. Beale MH, Hagan MT, Demuth HB.. *Neural Network Toolbox. User's Guide*, Vol. 2. Virgin Islands: MathWorks; 2010, pp. 77–81.

12. Shandilya P, Jain PK, Jain NK. RSM and ANN modeling approaches for predicting average cutting speed during WEDM of SiCp/6061 Al MMC. *Procedia Eng.* 2013;64:767–774.

13. Banerjee N, Biswas AR, Kumar M, Sen A, Maity SR. Modeling of laser welding of stainless steel using artificial neural networks. *Mater Today: Proc.* 2022.

14. Patel KA, Brahmbhatt PK. A comparative study of the RSM and ANN models for predicting surface roughness in roller burnishing. *Proc Technol.* 2016;23:391–397.

15. Paturi UMR, Cheruku S, Geereddy SR. Process modeling and parameter optimisationof surface coatings using artificial neural networks (ANNs): State-of-the-art review. *Mater Today: Proc.* 2021;38:2764–2774.

16. Guessasma S, Montavon G, Coddet C. On the neural network concept to describe the thermal spray deposition process: An introduction. In: Lugscheider E, Kammer PA, editors. *Proceedings of the International Thermal Spray Conference and Exposition.* Düsseldorf: DVS-Verlag GmbH; 2002. pp. 435–439.

17. Guessasma S, Salhi Z, Montavon G, Gougeon P, Coddet C. Artificial intelligence implementation in the APS process diagnostic. *Mater Sci Eng B.* 2004;110(3):285–295.

18. Ikeuchi D, Vargas-Uscategui A, Wu X, King PC. Data-efficient neural network for track profile modelling in cold spray additive manufacturing. *Appl Sci* 2021;11(4):1654.

19. Ikeuchi D, Vargas-Uscategui A, Wu X, King PC. Neural network modelling of track profile in cold spray additive manufacturing. *Materials.* 2019;12(17):2827.

20. Han BY, Xu WW, Zhou KB, Zhang HY, Lei WN, Cong MQ, Du W, Chu JJ, Zhu S. Performance analysis of plasma spray Ni60CuMo coatings on a ZL109 via a back propagation neural network model. *Surf Coat Technol.* 2022;433:128121.

21. Razavipour M, Legoux JG, Poirier D, Guerreiro B, Giallonardo JD, Jodoin B. Artificial neural networks approach for hardness prediction of copper cold spray laser heat treated coatings. J Therm Spray Technol. 2022; 31(3):525–544.

22. Mojena MAR, Roca AS, Zamora RS, Orozco MS, Fals HC, Lima CRC. Neural network analysis for erosive wear of hard coatings deposited by thermal spray: Influence of microstructure and mechanical properties. *Wear.* 2017;376:557–565.

23. Behera A, Mishra SC. Prediction and analysis of deposition efficiency of plasma spray coating using artificial intelligence method. *Open J. Comp Mater.* 2012; 2:54–60.

24. Lin CM, Yen SH, Su CY. Measurement and optimisationof atmospheric plasma sprayed CoMoCrSi coatings parameters on Ti-6Al-4V substrates affecting microstructural and properties using hybrid abductor induction mechanism. *Measurement.* 2016;94:157–167.

25. Szala M, Łatka L, Awtoniuk M, Winnicki M, Michalak M. Neural modelling of APS thermal spray process parameters for optimisingthe hardness, porosity and cavitation erosion resistance of Al2O3-13 wt% TiO2 coatings. *Processes* 2020;8(12):1544.

26. Kondo R, Yamakawa S, Masuoka Y, Tajima S, Asahi R. Microstructure recognition using convolutional neural networks for prediction of ionic conductivity in ceramics. *Acta Mater.* 2017;141:29–38.

27. Cecen A, Dai H, Yabansu YC, Kalidindi SR, Song L, Material structure-property linkages using three-dimensional convolutional neural networks. *Acta Mater.* 2018;146:76–84.

28. Yang Z, Yabansu YC, Al-Bahrani R, Liao WK, Choudhary AN, Kalidindi SR, Agrawal A. Deep learning approaches for mining structure-property linkages in high contrast composites from simulation datasets. *Comp Mater Sci.* 2018;151:278–287.

29. Lubbers N, Lookman T, Barros K. Inferring low-dimensional microstructure representations using convolutional neural networks. *Phys Rev E.* 2017;96(5):052111.

30. Ma B, Ban X, Huang H, Chen Y, Liu W, Zhi Y. Deep learning-based image segmentation for Al-La alloy microscopic images. *Symmetry*. 2018;10(4):107.
31. Azimi SM, Britz D, Engstler M, Fritz M, Mücklich F. Advanced steel microstructural classification by deep learning methods. *Sci Rep*. 2018;8(1):1–14.
32. Ling J, Hutchinson M, Antono E, DeCost B, Holm EA, Meredig B. Building data-driven models with microstructural images: Generalisationand interpretability. *Mater Discov*. 2017;10:19–28.
33. DeCost BL, Francis T, Holm EA. Exploring the microstructure manifold: Image texture representations applied to ultrahigh carbon steel microstructures. *Acta Mater*. 2017;133:30–40.
34. Li X, Zhang Y, Zhao H, Burkhart C, Brinson LC, Chen W. A transfer learning approach for microstructure reconstruction and structure-property predictions. *Sci Rep*. 2018;8(1):1–13.
35. Cang R, Xu Y, Chen S, Liu Y, Jiao Y, Yi Ren M. Microstructure representation and reconstruction of heterogeneous materials via deep belief network for computational material design. *J Mech Des*. 2017;139(7):071404.
36. Liu R, Wang M, Wang H, Chi J, Meng F, Liu L, Wang F. Recognition of NiCrAlY coating based on convolutional neural network. *NPJ Mater Degrad*. 2022;6(1):1–7.
37. Zhu J, Wang X, Kou L, Zheng L, Zhang H. Prediction of control parameters corresponding to in-flight particles in atmospheric plasma spray employing convolutional neural networks. *Surf Coat Technol*. 2020;394:125862.
38. Gebauer J, Gruber F, Holfeld W, Grählert W, Lasagni AF. Prediction of the quality of thermally sprayed copper coatings on laser-structured CFRP surfaces using hyperspectral imaging. *Photonics*92022; 9(7): 439.
39. Zhu J, Wang X, Kou L, Zheng L, Zhang H. Application of combined transfer learning and convolutional neural networks to optimize plasma spraying. *Appl Surf Sci*. 2021;563:150098.
40. Demirkiran AS, Celik E, Yargan M, Avci E. Oxidation behaviour of functionally gradient coatings including different composition of cermets. *Surf Coat. Technol*. 2001;142:551–556.
41. Usmani S, Sampath S. Time-dependent friction response of plasma-sprayed molybdenum. *Wear*. 1999;225:1131–1140.
42. Cetinel H, Öztürk H, Celik E, Karlık B. Artificial neural network-based prediction technique for wear loss quantities in Mo coatings. *Wear*. 2006;261(10):1064–1068.
43. *MATLAB User's Guide, Neural Network Toolbox*. Virgin Islands: The Math Works Inc; 2002.
44. Satapathy A, Mishra SC, Das R, Mishra SS, Ananthapadmanabhan PV, Sreekumar KP. Prediction of erosion behaviour of plasma sprayed fly ash coatings using neural network; In: DAE-BRNS Symposium on Power Beam Applications in Materials Processing PBAMP2006, Bhaba Atomic Research Centre, Mumbai (India) on 20th September 2006.
45. Pati PR, Satapathy A, Gupta G, Ray S. Optimizing wear analysis of plasma sprayed Linz-Donawitz slag-Al2O3 coatings using experimental design and neural network. *Proc Inst Mech Eng Part J: J Eng Tribol*. 2022;236(9):1723–1736.
46. Wang Z, Cai S, Chen W, Ali RA, Jin K. Analysis of critical velocity of cold spray based on machine learning method with feature selection. *J Therm Spray Technol*. 2021;30(5):1213–1225.
47. Wang L, Fang JC, Zhao ZY, Zeng HP. Application of backward propagation network for forecasting hardness and porosity of coatings by plasma spraying. *Surf Coat Technol*. 2007;201(9–11):5085–5089.
48. Jean MD, Lin BT, Chou JH. Application of an artificial neural network for simulating robust plasma-sprayed zirconia coatings. *J Am Ceramic Soc*. 2008;91(5):1539–1547.

49. Datta S, Pratihar DK, Bandyopadhyay PP. Modeling of input-output relationships for a plasma spray coating process using soft computing tools. *Appl Soft Comp.* 2012;12(11):3356–3368.

50. Pakseresht AH, Ghasali E, Nejati M, Shirvanimoghaddam K, Javadi AH, Teimouri R. Development empirical-intelligent relationship between plasma spray parameters and coating performance of Yttria-Stabilized Zirconia. *Int J Adv Manuf Technol.* 2015;76(5):1031–1045.

51. Guessasma S, Montavon G, Coddet C. Neural computation to predict in-flight particle characteristic dependences from processing parameters in the APS process. *J Therm Spray Technol.* 2004;13(4):570–585.

52. Guessasma S, Trifa FI, Montavon G, Coddet C. Al2O3-13% weight TiO2 deposit profiles as a function of the atmospheric plasma spraying processing parameters. *Mater Des.* 2004;25(4):307–315.

53. Kanta AF, Montavon G, Berndt CC, Planche MP, Coddet C. Intelligent system for prediction and control: Application in plasma spray process. *Expert Syst Appl.* 2011;38(1):260–271.

54. Liu T, Planche MP, Kanta AF, Deng S, Montavon G, Deng K, Ren ZM. Plasma spray process operating parameters optimisationbased on artificial intelligence. *Plasma Chem Plasma Process.* 2013;33(5):1025–1041.

55. Cetinel H, Öztürk H, Celik E, Karlık, B. Artificial neural network-based prediction technique for wear loss quantities in Mo coatings. *Wear.* 2006;261(10):1064–1068.

56. Zhang G, Kanta AF, Li WY, Liao H, Coddet C. Characterisationsof AMT-200 HVOF NiCrAlY coatings. *Mater Des.* 2009;30(3):622–627.

57. Liu M, Yu Z, Zhang Y, Wu H, Liao H, Deng S. Prediction and analysis of high velocity oxy fuel (HVOF) sprayed coating using artificial neural network. *Surf Coat Technol.* 2019;378:124988.

58. Martinez C, AsareYeboah F, Herford S, Brzezinski M, Puttagunta V. Predicting wind turbine blade erosion using machine learning. *SMU Data Sci Rev.* 2019;2(2):17.

59. Karlsson D. *Wind Turbine Performance Monitoring Using Artificial Neural Networks.* Masters Thesis, Department of Energy and Environment, Chalmers University of Technology, Gothenburg, Västra Götaland, Sweden 2015.

60. Yang AY, Cheng L. Two-step surface damage detection scheme using convolutional neural network and artificial neural network. In: *2020 IEEE 23rd International Conference on Information Fusion (FUSION).* IEEE; 2020. pp. 1–8.

61. Moreno-Armendáriz MA, Duchanoy CA, Calvo H, Ibarra-Ontiveros E, Salcedo-Castañeda JS, Ayala-Canseco M, García D. Wind booster optimization for on-site energy generation using vertical-axis wind turbines. *Sensors.* 2021; 21(14): 4775.

62. Xu D, Wen C, Liu J. Wind turbine blade surface inspection based on deep learning and UAV-taken images. *J Renew Sustain Energy.* 2019;11(5):053305.

63. Wang L, Zhang Z. Automatic detection of wind turbine blade surface cracks based on UAV-taken images. *IEEE Trans Ind Electron.* 2017;64(9):7293–7303.

64. Young A, Kay A, Marshall S, Torr R, Gray A. Hyperspectral imaging for erosion detection in wind turbine blades; In: Proceedings of HSI 2016, London, UK. 12–13th October 2016.

65. Rizk P, Al Saleh N, Younes R, Ilinca A, Khoder J. Hyperspectral imaging applied for the detection of wind turbine blade damage and icing. *Remote Sens Appl: Soc Environ.* 2020;18:100291.

# 3 Artificial Intelligence-Based Image-Processing Techniques for Assessment of Patterns and Mechanisms in Thermal Spray

## ABBREVIATIONS

| | |
|---|---|
| **AI** | Artificial intelligence |
| **CNN** | Convolutional neural network |
| **COV** | Coefficient of variance |
| **GAN** | Generative adversarial network |
| **IP** | Image processing |
| **I/P** | Input |
| **O/P** | Output |
| **SEM** | Scanning electron microscopy |
| **TCL** | Top coat layer |

## 3.1 INTRODUCTION TO IMAGE-PROCESSING TECHNIQUES

Image processing (IP) has been one of the fields that have greatly benefited from the advent of artificial intelligence (AI) [1–4]. The evaluation of wear patterns and wear processes in real time is a crucial application of AI in IP. Manufacturing, automotive, aviation, and healthcare are just a few of the fields where an understanding of wear patterns and wear processes is vital. Improving performance and decreasing maintenance costs may result from an accurate study of wear patterns and wear mechanisms, which can help detect possible concerns and minimize expensive downtime. AI-based IP systems can quickly and accurately analyze and detect wear patterns and wear processes, allowing for immediate input (I/P) on the condition of machinery and tools [5–9]. To analyze enormous volumes of data from pictures and uncover patterns that are typically difficult for people to perceive, AI systems employ deep learning, machine learning, and computer vision. There are several benefits of using AI in IP over more conventional methods. These methods can process massive volumes of data and provide instantaneous responses, making it simpler to spot problems before

DOI: 10.1201/9781003400660-3

they escalate. They are more precise than older techniques and can pick up on subtle changes in wear patterns that human analyzers would overlook. In addition, AI-based IP systems may be educated to recognize certain wear mechanisms, which can lead to the discovery of the underlying cause of the wear and the implementation of more precise maintenance. Several AI-based IP approaches are utilized to evaluate wear patterns and wear processes in real time. In this context, convolutional neural networks (CNNs) are a popular tool. To analyze wear, CNNs, a form of deep learning system, can learn characteristics from photos and spot patterns. When compared to more conventional techniques, they are superior at picking up on even the most imperceptible changes in wear patterns. Unsupervised machine learning is an additional method in which the algorithm discovers patterns in the data without being given any guidance on what to search for. When the wear mechanisms are unknown and the algorithm must automatically detect them, this method is invaluable. Generative adversarial networks (GANs) are another method being investigated for wear analysis [10]. GAN is an AI algorithm class that can create synthetic pictures from raw data. This method may be used to model wear processes and predict their future development.

## 3.2 VARIOUS IP TECHNIQUES

The term "image-processing techniques" describes a group of procedures and algorithms used for this purpose. There are three primary categories into which these methods fall:

Point processing methods are those that allow for the independent modification of picture pixels with no noticeable effect on their neighbors. Modifying the image's brightness, contrast, color, and gamma are all part of this process.

- One category of IP methods is called "neighborhood processing techniques," and it entails changing a cluster of pixels located in a certain area of the picture. Processes like noise cancellation, smoothing, and sharpening fall under this heading.
- Geometric processing techniques: These include operations on images that change their shape (e.g., rotation, scaling, and skewing). Image registration, fusion, and segmentation are three applications that often use geometric processing methods.

Other than these general methods, there are other application-specific methods, such as:

- Image form and structure may be analyzed and manipulated via morphological processing methods. Erosion, expansion, contraction, opening, and closure are all part of this process.
- Frequency domain processing approaches entail re-presenting a picture as a collection of frequency components, a transformation into the frequency domain. Transforms like the Fourier and wavelet transforms are part of this category.

As a result, there are primarily three classes of IP methods: point processing, neighborhood processing, and geometric processing. Applications like morphological processing, frequency domain processing, and compression need highly specialized methods. Each method has its own set of benefits and drawbacks, and it is chosen according to the needs of the given task.

## 3.3  IMAGING PROCESSING TECHNIQUES AND TOOLS

### 3.3.1  CONVOLUTIONAL NEURAL NETWORK

ANNs, which have progressed from Multilayer Perceptrons (MLP) [11] to the more recent deep CNN [12], are based on a hidden structure with a single output (O/P) and a multiple-layer configuration at the I/P. When I/P visual data is used, a CNN develops meaningful connections. It contains several interconnected levels, much like the human brain, which is reminiscent of the nervous system. Through activation functions like the sigmoid function, a signal from one layer of neurons may be translated into a response signal for the neurons of the next layer. There is a dynamic, nonlinear functional mapping between I/P and O/P [13]. If the filters are used effectively, CNN may learn spatial and temporal relationships. To apply a convolution operator [14], just place the appropriate filter on an image and multiply the resulting values from the filter and the picture. Images may be modified immediately by applying the filters. CNN can study and label photos [15]. Figure 3.1 depicts the use of CNN for glass-crack detection.

### 3.3.2  GENERATIVE ADVERSARIAL NETWORKS (GANS)

GANs are a kind of deep learning model that relies on generative modeling. The goal of this modeling technique is to deduce the underlying structure of the given data and reproduce it unsupervised [17]. These days, GANs are a supervised strategy that speeds up the training process. A generator used to transfer the random vectors into the produced samples and a discriminator are shown in Figure 3.2a, illustrating the architecture of MatGAN for inorganic materials [18]. To trick the discriminator, the generator fabricates training instances that are very similar to the I/P data. The discriminator's job is to determine whether or not the provided data are authentic. When the dataset size

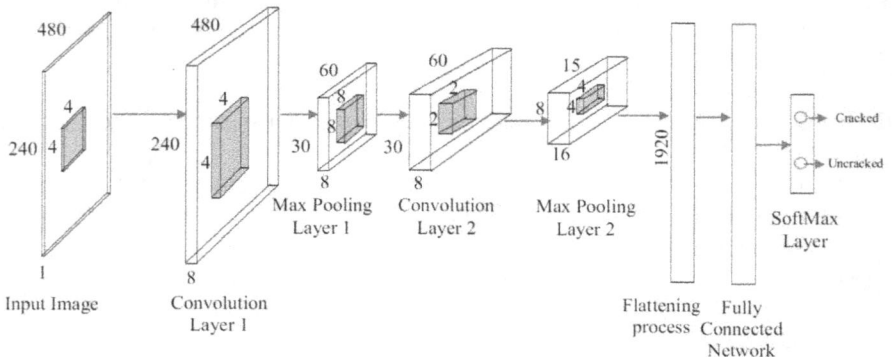

**FIGURE 3.1**  Crack analysis model of the convolutional neural network model [16]. Permissions under the CC-BY-NC-ND 4.0 license.

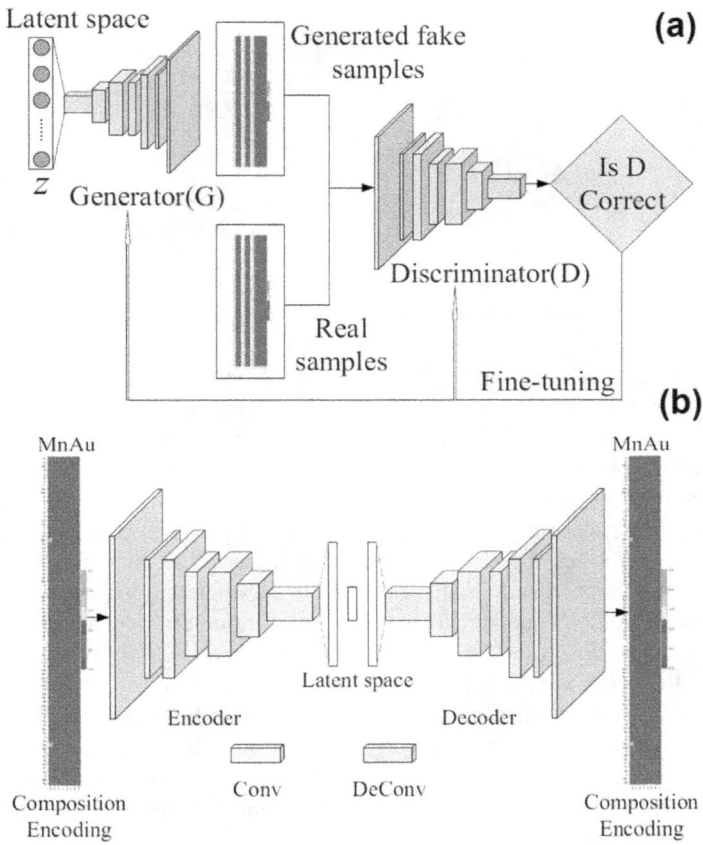

**FIGURE 3.2** Structures of (a) a materials-specific MatGAN and (b) a generative adversarial networks auto-encoder [18]. Permissions under CC Attribution 4.0.

is limited, it may be helpful. It's a common tool for making 3D models, adding effects to old photos, and enhancing the quality of existing ones. The goal of the discriminator is to tell the difference between natural and artificial materials. Convolutional and deconvolutional layers make up the encoder (Figure 3.2b). Encoders are the answer to the difficulty of determining the composition of materials from small samples [18].

## 3.4 IMPLEMENTATION OF A CONVOLUTIONAL NEURAL NETWORK (CNN) IN THERMAL SPRAY

To overcome the limitations of ANN models, a basic CNN model used by Zhu et al. [19] to predict control parameters during the plasma spray process is shown in Figure 3.3. The pooling layer's filter size is $4 \times 4$, while the convolutional layer's filter size is $6 \times 6$. The outcomes show the CNN models' capacity for generalization, which is useful for preparing target coatings' control parameters. CNN models

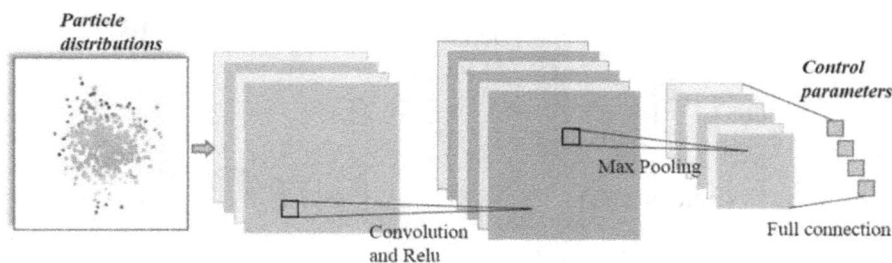

**FIGURE 3.3**   A basic convolutional neural network model [19]. Permissions from Elsevier.

are also significantly faster at predicting the outcome than traditional methods like numerical simulation. The simulation program took roughly 10 hours to generate the 3D distribution of the plasma jet after the control parameters were established. After training the CNN model, it took only 0.01 seconds to forecast the necessary control parameters from a set of in-flight particle characteristic distributions. As a result, using a neural network model to determine control parameters during the plasma spraying process saves time and computer resources. The back-propagation technique was used to learn and update the model's parameters (weight coefficients and biases) [20]. The CNN model includes several parameters that have been trained to reduce error or loss functions. The difference between CNN models' predicted values and the actual data is referred to as the loss function. The loss function, also known as the cost function, aims to inform us of how well the model is working, and the derivative of the loss function instructs us on how to adjust our parameters to improve the model's performance. There are numerous accessible loss functions, and the selection of a loss function depends on the problem being addressed. The mean square error is a frequently employed loss function for many optimization problems.

$$L = \frac{1}{n} \sum_{i=1}^{n} (\hat{y}_i - y_i)^2 \tag{3.1}$$

where $L$ is the loss, $n$ is the number of samples, $\hat{y}_i$ is the prediction, and $y_i$ is the true value.

The Adam algorithm [21] or stochastic gradient descent approach was frequently employed in the training of models to identify the best solution or minimal loss. The partial derivative of a group of parameters is the outcome of the common optimization technique known as gradient descent. The steeper the slope, the higher the gradient. Instead of using the complete data set, stochastic gradient descent derives the gradient from a randomly chosen subset. This lowers the computing cost for large optimization problems and speeds up iterations.

Additionally, some research has demonstrated that when used for microstructure quantification in thermal spraying coatings, CNNs can produce findings with satisfactory accuracy. To train CNN with different architectures, Chen et al. [22] randomly picked pixels from the top coat layer (TCL) to create sub-images that were centered at the pixel and had a range of sizes. On a dataset of 150 shots of size 100 by 100 randomly chosen from a batch of 30 high-resolution thermal barrier coating images, their suggested methodology was assessed. According to their findings, the CNN-based models had a reduced average relative error (ARE) of 0.113 and a higher average classification accuracy of 100% at the confidence level of 90% for a VGG16-based model.

Lu et al. [23] improved on the work of Chen et al. [22]. By including some new processes, such as data augmentation and transfer learning, an enhanced CNN approach was developed. The photos gathered for their investigation correspond to coatings made from three different types of powders: Type A: Metco 601NS, Type B: Metco 995C, and Type C: Metco 204BNS. The dataset consists of 159 raw shots, with 50, 49, and 60 Type A, Type B, and Type C images, respectively. The following seven phases can be used to summarize the complete training procedure suggested by Lu et al. [23]:

- By manually classifying each pixel in a raw image as one of the four classes—amount material, TCL—microstructure, TCL—coating material, or TCL—substrate, actual coating layer can be designed.
- Using the ground truth mask, extract the topcoat layer from the raw image.
- After selecting a sub-image size, randomly pick a pixel from the TCL, and use that pixel as the center to crop sub-images into that size.
- To create a training dataset, mirror and rotate the sub-images that were cropped.
- To train CNN models, resize the sub-images and feed them into CNNs with trained parameters.
- Choose the ideal CNN model.
- Applying the CNN model to sub-images taken from the TCL area will allow you to identify microstructure and coating material in the TCL area in a pixel-by-pixel manner, allowing you to gauge how well the CNN model performs in classifying microstructure and coating materials.

CNN was also employed in the modeling of the reverse process to explore the relationship between in-flight particle characteristics and control parameters, taking into account the 2-D feature of in-flight particle characteristics at a specific spraying distance [19]. Furthermore, a transfer learning method was suggested by Pan and Yang [25] to apply the knowledge acquired in the source task to the target task to solve the issues in scenarios with a small amount of data. In the study by Zhu et al. [24] a pretrained CNN model was used using a deep transfer learning technique, and the CNN model was transferred via various methods of fine-tuning to model the plasma spray reversal process with NiCrAlY. As shown in Figure 3.4, this study is divided into three main sections: preparation of data, deep transfer learning model, and validation of the model. According to the results, the technique that fine-tuned the entire pretrained CNN model while also slowing down the learning rate demonstrated the lowest loss in the training dataset and the maximum testing accuracy.

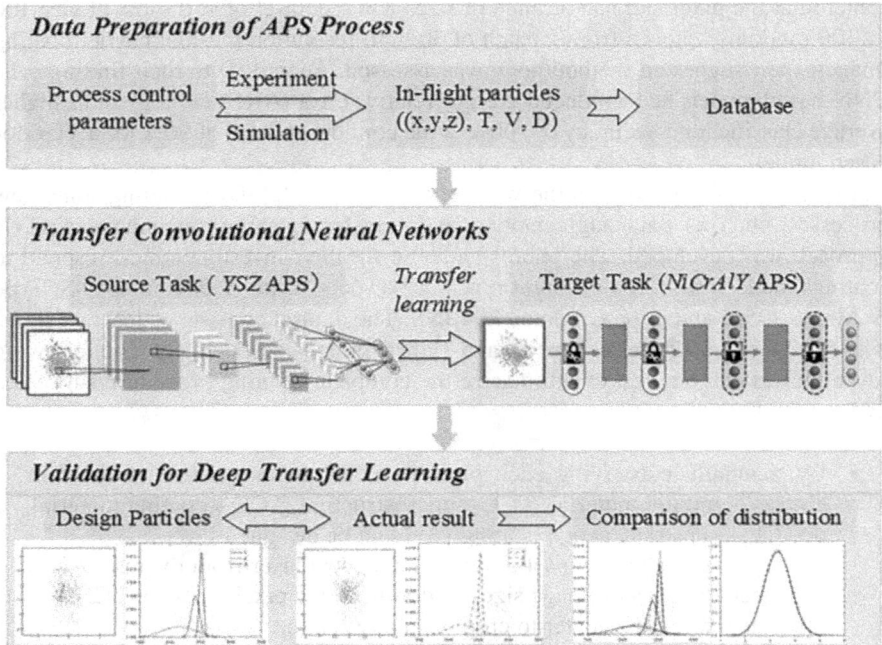

**Data Preparation of APS Process**

Process control parameters  ⟷ Experiment / Simulation ⟹  In-flight particles ((x,y,z), T, V, D)  ⟹  Database

**Transfer Convolutional Neural Networks**

Source Task ( YSZ APS )        Transfer learning        Target Task (NiCrAlY APS)

**Validation for Deep Transfer Learning**

Design Particles  ⟷  Actual result  ⟹  Comparison of distribution

**FIGURE 3.4**  Outline of deep transfer learning implementation in the reverse atomized plasma spray process [24]. Permissions from Elsevier.

## 3.5   CNNS IN PYTHON WITH KERAS

Deep learning is a field in which Keras is well-known. It is a pretty straightforward open-source Python library. It is simple to use. It is highly readable and has a very simple syntax. The importance of CNN increases when using deep learning to categorize images. The best Python library for handling CNN is Keras. Building a CNN is made simpler by this. The steps are briefly outlined below. CNNs are developed using Python (Version 3.7, with Navigator; Spyder/Anocoda):

- Gathering data is the first step. The adopted dataset might be the Fashion-MNIST dataset;
- The second stage entails importing libraries such as CNN, Keras, and Tensorflow;
- The dataset is split in the third phase using specific codes;
- Convolution, polling, and flattening are the final three sub-steps in the construction of the CNN model;
- The projected results can then be printed, exported, and examined after the CNN model has been trained and evaluated using image data.

## 3.6　CASE STUDY: IP FOR COATING-DEGRADING ERODENT CHARACTERIZATION

To characterize the coatings and eroding particles, researchers have utilized a wide range of interface instruments. Based on research. The physical parameters of the particles were measured with the use of an image-processing approach by Singh et al. [26]. They used an IP approach to evaluate the sphericity, solidity, circularity factor, and variance of particles. During scanning electron microscopy (SEM), experts advised using a magnification range of 500–2000 for the micrographs produced [27]. The particles in the SEM picture were highlighted using a threshold function and many other tools. The SEM pictures' threshold was modified using the Huang tool. The process of digital IP is shown in Figure 3.5. ImageJ was used to do the particle size analysis [28].

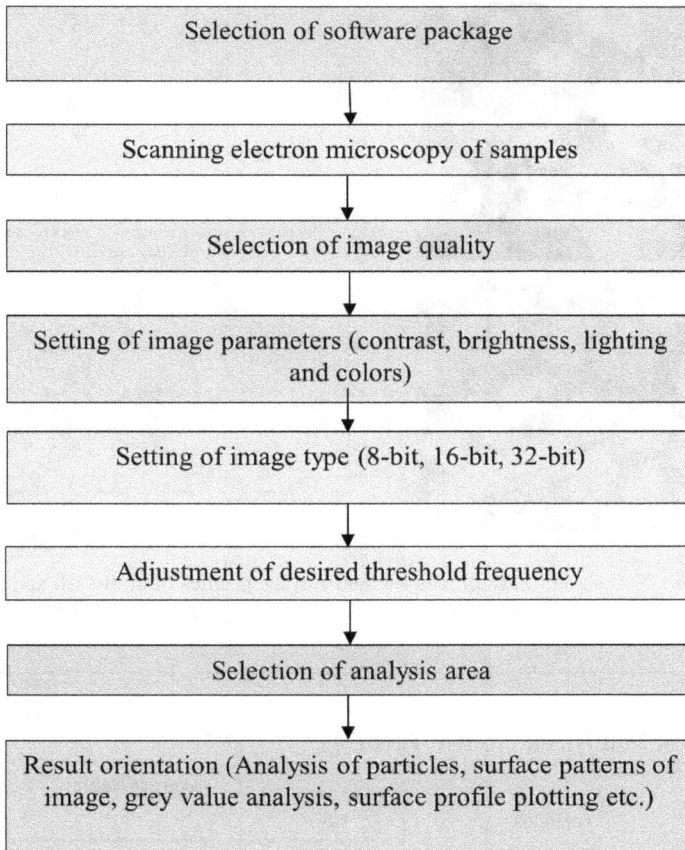

FIGURE 3.5　Schematic chart of the steps involved in the image processing of scanning electron microscopy micrographs.

### 3.6.1 Surface Smoothness Analysis

Figure 3.6 shows the scanning electron micrograph images that were analyzed to determine the form. The surface profiles of fly and bottom ash were created using the original SEM images as a basis. Fly ash particles have a smooth surface, which is reflected in the low waviness of the gray pattern curve formed for them. Bottom ash has an uneven surface, as seen by the waviness in its curvature.

### 3.6.2 Circularity (CF) Analysis

Table 3.1 is a summary of the attributes of the particles in the multisized slurry that was produced via the IP approach. Fly ash was found to have a CF of 0.908, which is very near 1. Walker and Humbe [27] state that spheres have a greater form factor

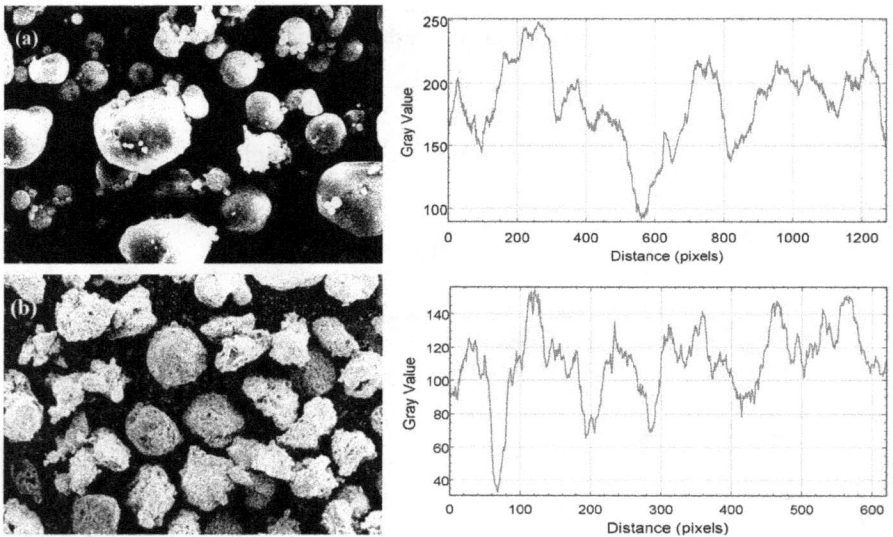

**FIGURE 3.6**  Threshold image processing and surface profiles of (a) fly ash and (b) bottom ash [30]. Permissions from Jashanpreet Singh.

---

**TABLE 3.1**
**CF Values of Multisized Erodent Particles**

| S. No. | Particles | Average Value | | |
| --- | --- | --- | --- | --- |
| | | CF | $\kappa$ | COV (%) |
| 1 | Fly ash | 0.908 | 0.917 | 5.68 |
| 2 | Bottom ash | 0.712 | 0.714 | 8.36 |
| 3 | Sand | 0.637 | 0.952 | 10.1 |

than angles do. Since this sand has sharp edges and triangular-shaped particles, the current experimental CF value of the sand is quite low (0.637) [29]. The ash pond's bottom ash was employed in the investigation [30]; it includes trace amounts of both unburned coal and fly ash. The bottom ash from these experiments had an average CF value of 0.712.

The varying CF values of the various solid particle forms are shown in Figure 3.7. Fly ash had a CF in the 0.75–0.95 range, bottom ash in the 0.55–0.80 range, and sand in the 0.5–0.75 range. The bulk of the SEM micrographs showed fly ash with CF values larger than 0.75, while a handful showed values less than 0.75, indicating aggregation of particles owing to a charge or moisture. The CF value for particles of equal size distribution was also investigated. Separating the particle size ranges of fly ash, bottom ash, and sand slurries allowed for the creation of particle size distribution of uniform size. Particle size distribution analyses on erodents were performed using British standard sieves. The CF value of uniform sludge, as determined by IP, is shown in Table 3.2.

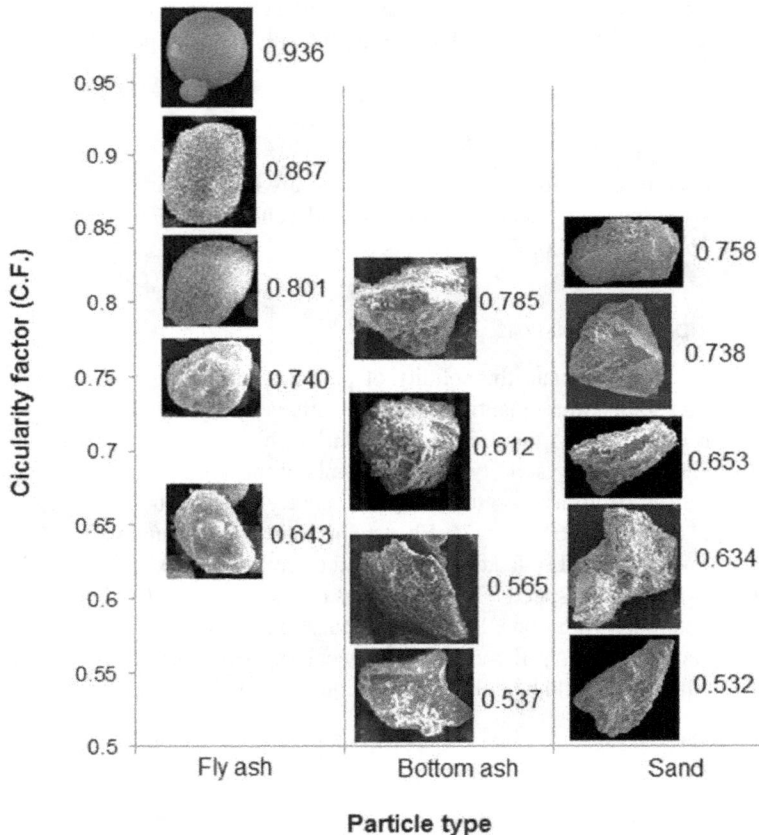

**FIGURE 3.7** Various types of particle shapes with CF values [30]. Permissions from Jashanpreet Singh.

**TABLE 3.2**

**CF of Equi-sized Erodent Media**

| S. No. | Particles | Particle Size Ranges | | | | |
|---|---|---|---|---|---|---|
| | | <53 μm | 53–75 μm | 75–106 μm | 106–150 μm | >150 μm |
| 1 | Bottom ash | 0.78 | | 0.61 | 0.56 | 0.54 |
| 2 | Sand | 0.76 | | 0.65 | 0.63 | 0.53 |
| 3 | Fly ash | 0.94 | 0.87 | 0.80 | 0.75 | 0.73 |

### 3.6.3 Sphericity ($\psi$) Analysis

During the study of the particles' images, the sphericity of various particle kinds was also evaluated. The IP software employed a ratio of sphere area to particle area to determine the size of random particles and calculate their perimeters [31]. The following correlation [32,33] was used to predict the sphericity of particles:

$$\psi = \frac{A_s}{A_p} \tag{3.2}$$

As in the preceding equation, $A_p$ is the particle area and $A_s$ is the sphere area. A random distribution of fly ash, bottom ash, and sand yielded sphericity values of 0.90–9.98, 0.78–0.89, and 0.64–0.76, respectively.

### 3.6.4 Solidity ($\kappa$) Analysis

IP was also used to evaluate the solidity of particles in this investigation. Particles' roughness and solidity were quantified using the image analysis tool's gray value measurements in a given SEM picture. The correlation that Liu et al. [34] analyzed successfully predicted particle solidity. Measurements of the solidity factors ($\kappa$) for evenly dispersed samples of fly ash, bottom ash, and sand yielded values between 0.90 and 9.98, 0.78 and 0.89, and 0.64 and 0.76. Fly ash, bottom ash, and sand were all randomly distributed, and their solidity factors were found to be between 0.89 and 0.94, 0.71 and 0.93, and 0.94 and 0.96, respectively. Sand seems to have a narrow range, but bottom ash particles have a much wider one. Fly ash continues to have a near-to 0.9 value. Solidity could not be completely considered in erosion wear experiments because the physically solid-surface material could be hollow inside, contain soft elements, be ductile, etc.

### 3.6.5 Effect of Image Environment

During simulations, image parameters might affect shape parameters. As can be seen in Figure 3.8, the CF value of the sand sample was determined using two distinct threshold procedures. Both parameters were found to provide distinct summaries of form parameters. The average CF was expected to be 0.733 and 0.837 in Figure 3.8a and b, respectively. This is because of the high circularity particles (shown by the

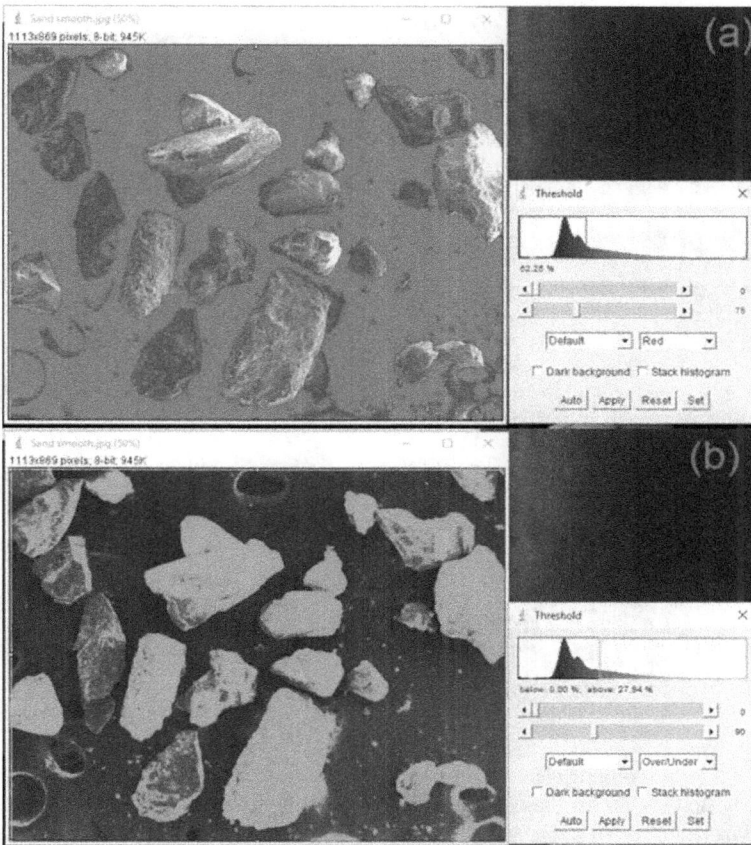

**FIGURE 3.8**    Application of threshold frequency [35]. Permissions from Elsevier.

dots in Figure 3.8) that are present. IP, it seems, provides the most accurate circularity measurement when the constraint factor is set to less than 0.99.

## 3.7    VARIOUS CHARACTERIZATION RESULTS OF THERMAL SPRAY COATINGS USING IMAGE PROCESSING

IP techniques are very effective in the characterization of erodent. Various characterizations like 3D topography, porosity assessment, 2D profilometry, roughness, cracks, etc. can be performed using the IP tools. Singh et al. [36] used DigitalSurf Mountains 8.0 [37] to test the 3D topography, porosity assessment, and 2D profilometry of WC-10Co4Cr coating, as shown in Figure 3.9. Imaging can be effective in testing erosion on the surface of coatings. Wear signatures can be easily tested using imaging tools. In a study, Singh et al. [38] performed IP to test the wear mechanisms on the surface of the Stellite 6 high-velocity oxy-fuel coating. They have found some underlying mechanisms like cracks, smear, and plowing to understand the phenomenon of wear, as shown in Figure 3.10.

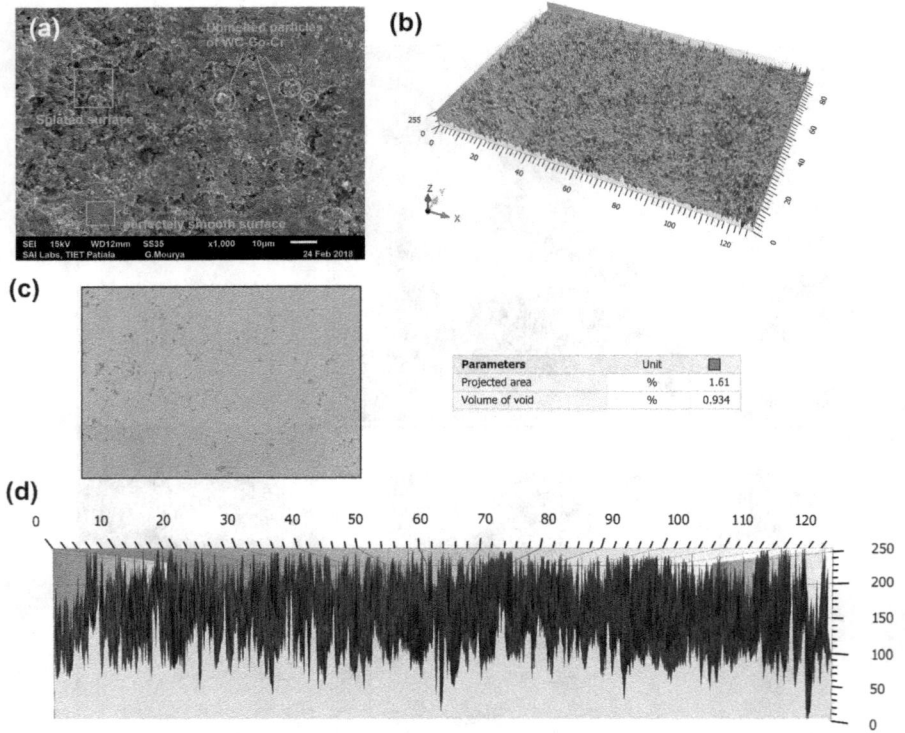

**FIGURE 3.9** (a) Scanning electron microscopy micrographs, (b) 3D topography, (c) porosity assessment, and (d) 2D profilometry of WC–10Co4Cr high-velocity oxy-fuel coatings [36]. Permissions from Elsevier.

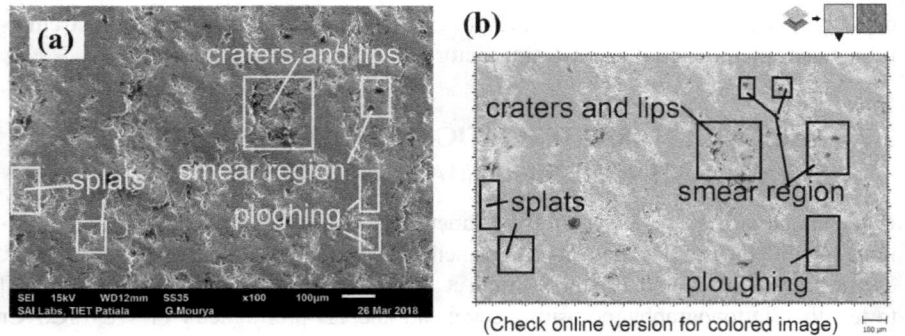

**FIGURE 3.10** (a) Scanning electron microscopy micrographs and (b) 3D topography of erosion mechanisms on Stellite 6 high-velocity oxy-fuel coatings [38]. Permissions from Elsevier.

## 3.8 CONCLUSIONS AND FUTURE PERSPECTIVE

AI-based IP algorithms have completely changed wear analysis by providing instantaneous data. These approaches are more precise than the ones that have been used in the past, and they can pick up on minute variations in wear patterns. As AI develops further, these methods will probably become even more refined and play a larger role in avoiding breakdowns, maximizing output, and decreasing upkeep expenses. IP methods that are based on AI help analyze the characteristics of solid particles, such as their size, shape, and symmetry, as well as their density. Image-processing techniques are helpful when attempting to draw 3D surface plots as well as 2D plots for roughness tests, patterns, wear processes, and porosity, among other things. Through the use of hyperspectral imaging technology, it is possible to quantify the material surfaces and wear rates of both coated and uncoated wind turbine blades.

## REFERENCES

[1] Singh S, Kasana SS. Estimation of soil organic carbon for sustainable agriculture using deep learning. *Proc 16th Int Conf Environ Sci Technol*. 2022;16:1–2.
[2] Singh S, Kasana SS. Quantification of soil properties from hyperspectral data for sustainable agriculture using deep learning. *Proc 16th Int Conf Environ Sci Technol*. 2022;16:2–3.
[3] Singh S, Kasana SS. Quantitative estimation of soil properties using hybrid features and RNN variants. *Chemosphere* [Internet]. 2022;287:131889. Available from: https://doi.org/10.1016/j.chemosphere.2021.131889.
[4] Singh S, Singh D, Sajwan M, et al. Hyperspectral image classification using multiobjective optimization. *Multimed Tools Appl*. 2022;81:25345–25362.
[5] Singh S, Kasana SS. A pre-processing framework for spectral classification of hyperspectral images. *Multimed Tools Appl*. 2021;80:243–261.
[6] Singh S, Kasana SS. Estimation of soil properties from the EU spectral library using long short-term memory networks. *Geoderma Reg* [Internet]. 2019;18:e00233. Available from: https://doi.org/10.1016/j.geodrs.2019.e00233.
[7] Singh S, Kasana SS. Spectral-spatial hyperspectral image classification using deep learning. *Proc 2019 Amity Int Conf Artif Intell AICAI 2019*. 2019;411–417.
[8] Singh S, Kasana SS. Efficient classification of the hyperspectral images using deep learning. *Multimed Tools Appl*. 2018;77:27061–27074.
[9] Singh S, Kasana SS. *Brief Review of Deep Learning Applications*, pp. 1–6.
[10] Singh J, Singh S. A review on machine learning aspect in physics and mechanics of glasses. *Mater Sci Eng B* [Internet]. 2022;284:115858. Available from: https://doi.org/10.1016/j.mseb.2022.115858.
[11] Camacho Olmedo MT, Paegelow M, Mas JF, et al. *Geomatic Approaches for Modeling Land Change Scenarios. An Introduction*. Cham, Switzerland: Springer Nature; 2018.
[12] Bai C, Huang L, Pan X, et al. Optimization of deep convolutional neural network for large scale image retrieval. *Neurocomputing* [Internet]. 2018;303:60–67. Available from: https://doi.org/10.1016/j.neucom.2018.04.034.
[13] Horne BG. Progress in supervised neural networks. *Signal Process Mag IEEE*. 1993;10:8–39.
[14] Kuo CCJ. Understanding convolutional neural networks with a mathematical model. *J Vis Commun Image Represent*. 2016;41:406–413.

[15] Spanhol FA, Oliveira LS, Petitjean C, et al. Breast cancer histopathological Im1. In: Spanhol FA, Oliveira LS, Petitjean C, Heutte L, editors. Breast Cancer Histopathological Image Classification Using Convolutional Neural Networks. *Int Jt Conf Neural Networks (IJCNN 2016)* ; 2017. pp. 1868–1873.

[16] Kouzehgar M, Krishnasamy Tamilselvam Y, Vega Heredia M, et al. Self-reconfigurable façade-cleaning robot equipped with deep-learning-based crack detection based on convolutional neural networks. *Autom Constr* [Internet]. 2019;108:102959. Available from: https://doi.org/10.1016/j.autcon.2019.102959.

[17] Goodfellow I, Pouget-Abadie J, Mirza M, et al. Generative adversarial networks. *Commun ACM*. 2020;63:139–144.

[18] Dan Y, Zhao Y. Generative adversarial networks (GAN) based efficient sampling of chemical composition space for inverse design of inorganic materials. *NPJ Comput Mater* [Internet]. 2020;6:1–7. Available from: https://doi.org/10.1038/s41524-020-00352-0.

[19] Zhu J, Wang X, Kou L, et al. Prediction of control parameters corresponding to in-flight particles in atmospheric plasma spray employing convolutional neural networks. *Surf Coat Technol*. 2020;394:125862.

[20] Rumelhart DE, Hinton GE, Williams RJ. Learning representations by back-propagating errors. *Nature*. 1986;323:533–536.

[21] Kingma DP, Ba J. Adam: A method for stochastic optimization. arXiv Prepr arXiv14126980; 2014.

[22] Chen W-B, Standfield BN, Gao S, et al. A fully automated porosity measure for thermal barrier coating images. *Int J Multimed Data Eng Manag*. 2018;9:40–58.

[23] Lu Y, Chen W-B, Wang X, et al. Deep learning-based models for porosity measurement in thermal barrier coating images. *Int J Multimed Data Eng Manag*. 2020;11:20–35.

[24] Zhu J, Wang X, Kou L, et al. Application of combined transfer learning and convolutional neural networks to optimize plasma spraying. *Appl Surf Sci*. 2021;563:150098.

[25] Pan SJ, Yang Q. A survey on transfer learning. *IEEE Trans Knowl. Data Eng*. 2009;22:1345.

[26] Singh J, Kumar S, Mohapatra S. Study on role of particle shape in erosion wear of austenitic steel using image processing analysis technique. *Proc Inst Mech Eng Part J J Eng Tribol* [Internet]. 2019;233:712–725. Available from: https://journals.sagepub.com/doi/10.1177/1350650118794698.

[27] Walker CI, Hambe M. Influence of particle shape on slurry wear of white iron. *Wear* [Internet]. 2015;332–333:1021–1027. Available from: https://doi.org/10.1016/j.wear.2014.12.029.

[28] Rasband WS. *ImageJ* [Internet]. Bethesda, MD: U.S. National Institutes of Health; 1997. Available from: https://imagej.nih.gov/ij.

[29] Woldman M, van der Heide E, Schipper DJ, et al. Investigating the influence of sand particle properties on abrasive wear behaviour. *Wear* [Internet]. 2012;294–295:419–426. Available from: https://doi.org/10.1016/j.wear.2012.07.017.

[30] Singh J. *Investigation on Slurry Erosion of Different Pumping Materials and Coatings.* Patiala, India: Thapar Institute of Engineering and Technology; 2019.

[31] Trahan J, Graziani A, Goswami DY. Evaluation of pressure drop and particle sphericity for an air-rock bed thermal energy storage system. *Energy Procedia*. 2014;57:633–642.

[32] Wong CY, Solnordal C, Swallow A, et al. Predicting the material loss around a hole due to sand erosion. *Wear* [Internet]. 2012;276–277:1–15. Available from: https://doi.org/10.1016/j.wear.2011.11.005.

[33] Bouwman AM, Bosma JC, Vonk P, et al. Which shape factor(s) best describe granules? *Powder Technol*. 2004;146:66–72.

[34] Liu T, Luo XT, Chen X, et al. Morphology and size evolution of interlamellar two-dimensional pores in plasma-sprayed La2Zr2O7 coatings during thermal exposure at 1300°C. *J Thermal Spray Technol* [Internet]. 2015;24:739–748. Available from: https://doi.org/10.1007/s11666-015-0236-0.

[35] Singh J, Kumar S, Mohapatra SK, et al. Shape simulation of solid particles by digital interpretations of scanning electron micrographs using IPA technique. *Mater Today Proc* [Internet]. 2018;5:17786–17791. Available from: https://doi.org/10.1016/j.matpr.2018.06.103.

[36] Singh J, Kumar S, Mohapatra SK. Tribological performance of Yttrium (III) and Zirconium (IV) ceramics reinforced WC-10Co4Cr cermet powder HVOF thermally sprayed on X2CrNiMo-17-12-2 steel. *Ceram Int* [Internet]. 2019;45:23126–23142. Available from: https://doi.org/10.1016/j.ceramint.2019.08.007.

[37] DigitalSurf. MountainsLab premium 8.0 Software: Trail Version [Internet]. 2020. Available from: https://www.digitalsurf.com.

[38] Singh J, Kumar S, Mohapatra SK. Erosion tribo-performance of HVOF deposited Stellite-6 and Colmonoy-88 micron layers on SS-316L. *Tribol Int* [Internet]. 2020;147:105262. Available from: https://doi.org/10.1016/j.triboint.2018.06.004.

# 4 Artificial Intelligence and Automation in Sustainable Development

## ABBREVIATIONS

| | |
|---|---|
| **AI** | Artificial intelligence |
| **APS** | Atmospheric plasma spray |
| **BC** | Bond coat |
| **CIC** | Chemical index of change |
| **CNN** | Convolutional neural network |
| **I/P** | Input |
| **ML** | Machine learning |
| **O/P** | Output |
| **TBC** | Thermal barrier coating |
| **TGO** | Thermally grown oxides |
| **TS** | Thermal spraying |
| **YSZ** | Yttria-stabilized zirconia |

## 4.1 INTRODUCTION TO SUSTAINABILITY

It is well known that the oil and gas and aerospace sectors account for the vast majority of the thermal spraying market. Thermal spray coatings are frequently used to boost a component's efficiency by increasing its resistance to wear, corrosion, and high temperatures. Using a variety of coatings, we can make parts last longer and make them more resistant to the environment they face. In addition to their use in the manufacturing of new parts, thermal spray coatings see widespread use in the refurbishment of older parts. Over a century has passed since thermal spray coatings were first used. Plasma spray (PS), detonation guns, high-velocity oxy-fuel, cold spray, and suspension plasma spray coating technologies are all examples of how this method has progressed over the years [1,2]. Thermal spray robots first appeared in manufacturing just before the turn of the century. Digital transformation and artificial intelligence (AI) represent the next step in technology development in this area [3].

## 4.2 AI AND SUSTAINABLE DEVELOPMENT (THERMAL SPRAY DIGITALIZATION)

The ecological effects of every manufacturing operation will take priority over its economic components. This is because climate change is having an overall effect on the economy. When it comes to the effects of climate change on national economies,

 DOI: 10.1201/9781003400660-4

governments will shift their focus from individual sectors to the economy as a whole. Thermal spray requires raw ingredients and produces waste products much like any other manufacturing process. Since the current trend is toward a more circular economy, maximizing sustainability in terms of the triple bottom line (environment, economy, and society) necessitates quantifying and tracking resources and waste through a life cycle sustainability assessment [4]. Typically, the efficiency of powder deposition is between 50% and 60%, meaning that 40% of the material is sent to the dump or recycling units. Some high-temperature gas turbines require segmented thermal barrier coatings (TBCs), which can be achieved via the solution precursor method or suspension precursor approach [5]. While powder techniques can increase output, they require a high-energy procedure, whereas spraying a solution or suspension decreases output but uses much less energy. Masks made of polymer or resin are considered solid waste and must be disposed of in a landfill. Therefore, innovations in ecologically friendly masks are needed to sustain the thermal spray sector. The top five industries that use thermal spray are depicted in Figure 4.1 as the largest emitters of greenhouse gases (GHGs) such as $CH_4$, $CO_2$, and $NiO_2$. Emissions from aircraft and power plants primarily come from the plasma spray process. Emissions come from the combustion of hydrocarbons used to coat drilling equipment in the oil and gas industry. Hence OEMs and service providers in the thermal spray industry should adopt a "culture of durability" that prioritizes reusing and recycling materials whenever possible to reduce waste.

The proliferation of AI is having an impact on a wide variety of industries. For example, it is anticipated that AI will have both immediate and long-term implications for the

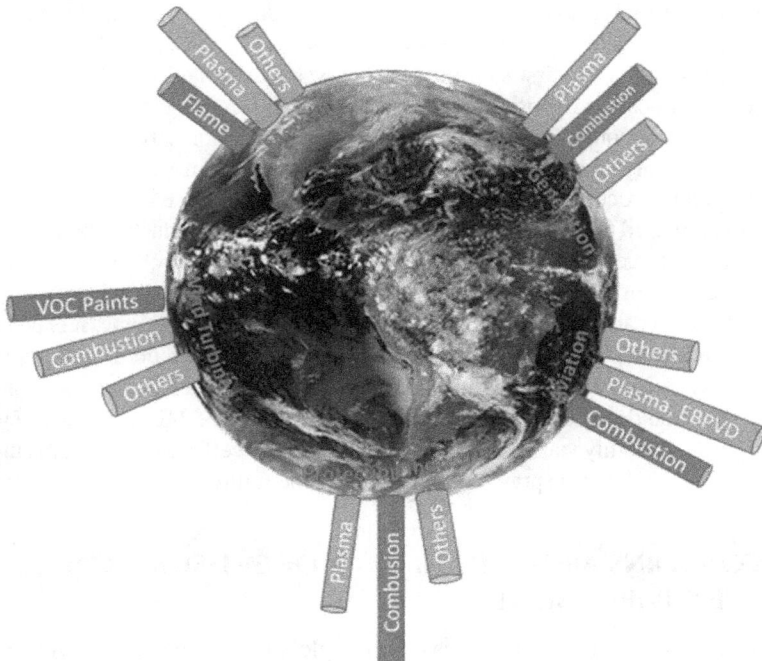

**FIGURE 4.1** Process and contribution of different industries to greenhouse gas emissions [5]. Permissions under the CC BY 4.0 license.

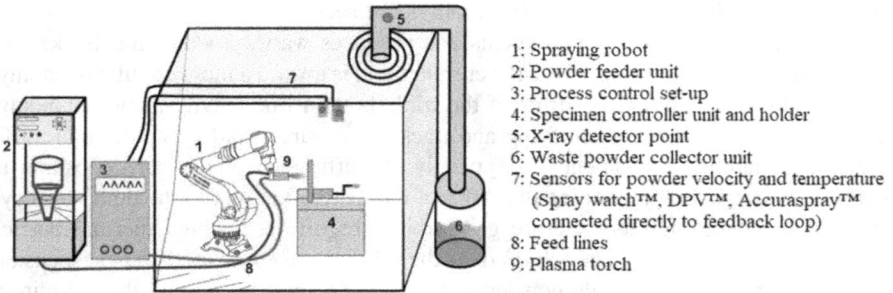

1: Spraying robot
2: Powder feeder unit
3: Process control set-up
4: Specimen controller unit and holder
5: X-ray detector point
6: Waste powder collector unit
7: Sensors for powder velocity and temperature
    (Spray watch™, DPV™, Accuraspray™
    connected directly to feedback loop)
8: Feed lines
9: Plasma torch

**FIGURE 4.2**  Schematic diagram of a sustainable thermal spraying unit (data gathering sites for digital I/P that lead to digital O/P and then on to the spraying robot and guns are shown by red dots) [5]. Permissions under the CC BY 4.0 license.

overall productivity of the world's workforce [6], equality and inclusion [7], environmental outcomes [8], and many other domains [9]. Thermal spray can be a part of the shift toward using AI and machine learning (ML) [10] to decrease emissions and waste. The potential for closed-loop feedback, in-situ stress measurement, and robot interfacing to be incorporated by original equipment manufacturers into intelligent thermal spray machines is yet to be determined. Governments may soon require thermal spray firms to provide emissions and waste data for analysis, making data analytics a crucial component of the process. The collection and analysis of data may be an integrated aspect of the thermal spray plant. Common in both industry and academia, the thermal spray arrangement depicted in Figure 4.2 includes several sensors and a waste collector. Data gathering is the first step in reducing waste and making the switch to more energy-efficient coating deposition procedures. Figure 4.2 suggests that sensors could be installed at key points inside the booth to gather this information. (i) Process data, like particle velocity and temperature, can be collected to optimize the gun parameters, robot speed, and process gas flows in real time, and (ii) resource use and waste can be tracked and reported to regulatory agencies. To better monitor industry compliance with climate change targets, we can classify them as "green," "yellow," or "red" according to the percentage of waste produced in each category. Through cloud computing, the accumulated data can be connected to servers and retrieved in real time. It is possible to construct mathematical models for optimizing the fuel/oxygen ratio in any combustion spray process to identify the optimal mix to melt the particles, lower fuel consumption, and increase deposition efficiency [11]. With the help of ML, robot coating pathways might be smoothly learned and modified to optimize efficiency [12]. This unit will potentially impact thermal spray on digitization in the future.

## 4.3   CONCERNS ABOUT THE IMPACT OF EMISSIONS ON THE ENVIRONMENT

Less than 5% annual growth is forecast for the global aviation industry over the next two decades. Current market trends were used to create this outlook. Population growth affects the environment by causing climate change, increased noise, and lower air quality. These shifts are inevitable because of the greater demand for goods and services that will emerge from the population boom. Possible mitigation strategies include setting laws and standards, improving aircraft and engine performance and/or developing alternative fuels, enhancing operations, and implementing market-based policies [13]. Flight frequency reduction, fuel efficiency enhancement, and the creation of substitute fuels are a few more potential mitigating strategies. Transportation demand may be affected in a variety of ways on a global and regional scale as a result of climate change. Anytime now, these alterations could be implemented. As temperatures rise, especially with the predicted increases in temperature, it is possible that the performance of airplanes and, consequently, the travel patterns of passengers will be greatly altered [14]. The Intergovernmental Panel on Climate Change concluded in its most recent assessment that most of the observed increases in global temperature can be attributed to human activities with a high degree of certainty. The Intergovernmental Panel on Climate Change presented this finding as part of its Fifth Assessment Report. The aviation sector is blamed for approximately 2% of global $CO_2$ emissions. The aviation industry has publicly committed to achieving these three ambitious climate goals. One of the goals is to maintain 2020 levels of net aviation $CO_2$ emissions, and to do this, a global market-based measure [15] has been established. Not only does aviation contribute to GHG emissions, but soot and sulfate particles released during combustion may also have an impact on the climate. All three of these factors affect the amount of GHGs in the air, which could have a knock-on effect on the weather.

It is widely agreed upon that effects that are not caused by $CO_2$ contribute to the warming of the globe as a whole, even though some additional effects assist in warming while others contribute to cooling. This is still the case even when certain supplementary consequences help speed up warming and others speed up cooling. When describing the climate impacts of aviation from non-$CO_2$-producing sources, a multiplier is sometimes utilized. This metric calculates the fractional influence of aviation on the environment by dividing the total effect of $CO_2$ on the environment [16]. To get this metric, you divide aviation's total climate impact by $CO_2$'s climate impact. Power output, thermal efficiency, and fuel economy are all areas in which internal combustion engines excel, yet the combustion process within these engines produces pollutants that are bad for the environment. Carbon monoxide, nitrogen oxides, and hydrocarbons are only a few examples of these noxious emissions. Today's internal combustion engines are well recognized as a major source of pollution. This is because these engines emit a potentially lethal cocktail of gases and particles into the air when they are running. As a result, many people face a wide range of challenges. Because of this, researchers and engineers have been hard at work on a new version of the engine that generates fewer harmful emissions. TBCs can help the internal combustion engines in cars by increasing the intake air temperature [17]. The ride will be more relaxing as a result of this. This aids in achieving high efficiency, extending the life of the engine's components, and reducing the emission of harmful compounds. Studies looking into various methods for reducing the aforementioned pollutants. Refractory metal oxides are applied to a substrate, and then a ceramic coating is applied on top. This type of ceramic coating is then put on components to prevent heat from escaping as the components operate at high temperatures. TBCs are composite overlays used on superalloy surfaces. Both a bond coat and a ceramic coat make up these layers of protection. This type of coating is commonly used in internal combustion engines. Tolerances must be applied appropriately to structural compliance to account for the thermal and mechanical strains executed by the exposure of services [18]. Adding brittle components to ductile substrates in the layering order mentioned above simplifies the process. Intergranular coating networks become twisted when methods like plasma spray and electron beam physical vapor deposition are performed side-by-side. Using this connected system, the aforementioned objective is realizable.

In contrast to cementation or continuous section thickness, porous deposition is typically used. There is no way around applying a TBC to the parts of an aircraft engine that are subjected to stress-limiting conditions [19]. This is because the weather has been getting colder and windier recently. The subsequent advantages include keeping high-temperature components safe so they may function at their best within their allowed temperature ranges and making the most efficient use of available energy. The solution's execution led directly to these two advantages. In-situ alumina production from substrate/bond coat aluminum and the transformation of metastable tetragonal zirconia into stable tetragonal zirconia, respectively, occur at the temperature of service, enhancing the thermo-mechanical behavior of TBC [20]. There is no difference in temperature between the two operations. In contrast to the latter, which leads to a reduction in volume, the former causes an expansion of volume. With today's composite technology, it's possible to give one's inventions a self-toughening effect. The unusual mechanical behavior of TBC has been linked to the presence of a tortuous network between the grains [21]. The metastable tetragonal phase transforms into the cubic allotropic state during the second stage of fracture tip blunting. This change is now underway. This relates to the stress that can be observed in the cracks. As a result, the Chemical Index of Change (CIC) has decreased. The CIC quotient will lead to an improvement in the material's toughness. Volumetric expansion pressures caused by extended exposure, however, will lead to nickel enhancement of thermally produced oxides (TBC) at the bonding coat/ceramic coat interface, which will lead to localized spallation zones. Pressures are accumulating at the bond coat/ceramic coat interface due to nickel enrichment at the interface. When TBC comes into contact with the substrate, it significantly alters the ductility of the material. Applying a bond coat has several benefits, including stress reduction and improved mechanical adherence. When it comes to bond coating, the M-CrAlY family of alloys is by far the most common choice for a wide variety of applications. Contacts that are heated to operating test temperatures result in the formation of thermally grown oxides (TGO). The TGO generation temperature refers to this heat level. As a direct outcome of interactions between the qualities, this occupies a middle ground. The basic components of TGO [22] are alumina that has been coupled to yttria and chromia and that has been catalyzed by chromia. To alleviate the adverse consequences of thermal expansion mismatch stresses, more study into the rivalry among auto-sintering and auto-toughening mechanisms is required, as is the use of composite coatings crafted from functionally graded materials that are microlaminated and multilayered in ceramic/ceramic [23]. Microlaminated, multilayered ceramics/ceramics with the appropriate functional gradations can achieve this goal. Microlaminated, functionally graded materials, in addition to multilayered ceramics, will be used to achieve these goals.

## 4.4  ECONOMIC AND ECOLOGICAL ASPECTS OF THERMAL SPRAY

Both the physical world and the abundance of available information are undergoing dramatic changes in the twenty-first century. Increased consumer demand and stricter regulations concerning the environment are placing increased pressure on businesses to improve the global quality of their products while simultaneously minimizing the damage to the natural world and the climate [24]. This is because strict environmental rules and higher expectations from consumers have contributed to this trend. Increasing production while decreasing costs and minimizing damage to the environment is of paramount importance. In addition to acting as an insulator, the yttria-stabilized zirconia (YSZ) coating serves to decrease the transfer of heat into the turbine's metal parts, which is particularly useful in high-temperature environments [25]. The metallic bond coat (BC) is a layer that protects the BC from the elements by covering it with oxidation- and corrosion-resistant metal. This is achieved by enhancing the bond between the component and the ceramic topcoat, which in turn protects the underlying component. Atmospheric plasma spray (APS) helps treat the engine components with YSZ. Due to the use of old torches, deposition efficiency levels remain low [26]. Thus, at least half of the YSZ sprayed did not settle on the components. It is crucial to stress that the YSZ powder that is used in the thermal spray process cannot be reused in any other way. It's hard to overstate how important it is to stress this. It is critical to study the problem and develop strategies for either decreasing the consumption of masking materials or finding new uses for them. This is because masks and other polymeric substances may be harmful to the ecosystem if used improperly. Furthermore, thermal spray recycling initiatives have been attempted multiple times [27]. It has been shown that the other uses for ceramic powders recovered from dust collectors are, at best, "timid." Because of this factor, many in the business believe that rates of deposition efficiency lower than 50% may soon be disregarded. For the company, this represents a huge opportunity for growth.

Modern processing methods allow even the oldest APS torches to be controlled by computers. Mass-flow meters are used as a crucial instrument in the process of establishing and maintaining stable plasma gas levels. Furthermore, the pace at which the torch is losing heat is monitored. Previous APS torches used copper nozzles lined with tungsten, and they were renowned for their outstanding efficiency. Tungsten-lined nozzles reduce TBC contamination during spraying [28–30] and last longer than ordinary nozzles. This innovation directly enables processing adjustments to be rapidly addressed before the completion of spraying. Figure 4.3 is a diagram of the thermal spray process's inputs and outputs.

Many diverse materials are used in today's world for a wide range of purposes, from those that require biomaterials or high temperatures. The main goal of feedstock materials is to boost functionality in extreme conditions. $Al_2O_3$ is the chemical formula for aluminum oxide; however, alumina is the more popular term. Many

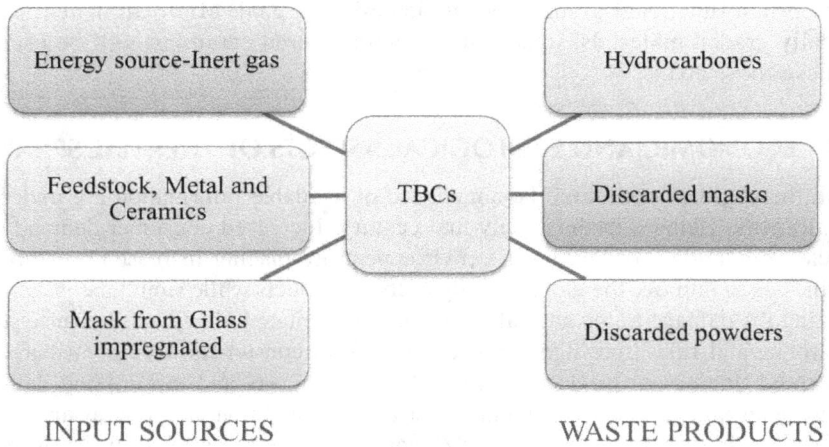

**FIGURE 4.3** Inputs and outputs in thermal spray.

boilers, gas turbine blades [31], and other high-temperature components used in medical devices and prostheses are made from alumina. It has shown excellent corrosion resistance in laboratory tests with living organisms. There are a growing number of situations where a certain type of material is used to survive challenging circumstances, similar to the application of YSZ in TBC [32]. The advancement of TBCs has made it possible for future gas turbines to run at significantly higher temperatures. This has led to a considerable amount of research being done to find novel materials that can outperform YSZ, the current industry standard. This is because YSZ is the material with the highest reputation in its sector [33]. It is suggested to use an integrated method of experiments and rational explanations after looking at previous advances and offering an explanation based on them. After reviewing the most recent developments, we propose the following. Since gas turbine engines are subjected to such a wide variety of operating circumstances, it has been challenging to design and develop the most efficient components for these engines during the past few decades [34]. Superalloys, including nickel, chromium, cobalt, titanium, and silver, were once often used in the manufacturing of gas turbine parts. Niobium, chromium, cobalt, and titanium alloys are also superalloys. Superalloys based on nickel are the most effective materials to work with for turbine part production [35–37]. Gas turbines are used in environments that are both hot and abrasive, and in recent years, research has been conducted across a wide range of disciplines to improve their performance in these conditions. The findings of this study will be used to make gas turbines more effective. A better-performing gas turbine will last longer, be more efficient while running, and release fewer pollutants. The most effective method of meeting these criteria is to increase the maximum operating temperature of the turbine, which will increase the device's thermodynamic efficiency [38].

To improve the turbine's performance and extend its service life, it's crucial to employ the right material. Several types of stainless steel were used in the early development of gas turbine engines; however, this did not improve their heat resistance. Nickel-based superalloys, such as Inconel, have largely supplanted stainless steels as the material of choice. The exceptional resistance to heat exhibited by nickel-based superalloys [39–41] is largely responsible for this. Single-crystal (SC) coating is a unique coating technology developed as a substitute for conventional coating practices for the production of single-crystal blades. When compared to directed solidification blades, single-crystal blades can function at higher temperatures. Crucial parts, however, can be damaged by direct exposure to extremely high temperatures without any thermal shielding [42–44]. The components will be harmed as a result of this. The TBC is used to cover the gas turbine components exposed to high-temperature conditions. The blades of a gas turbine can be cooled by encasing them in a material with low thermal conductivity. This will ensure the components can be used without risk. The gas turbine benefits from this since it can operate more effectively [45–47]. Because TBCs reduce the operating temperature of metal parts, they can be used for longer without degrading performance.

## 4.5 AVIATION EMISSIONS

The impact of jet aircraft emissions on the environment is shown in the diagram to the right (Figure 4.4). When jet fuel containing kerosene is burned, a cloud of fumes and particles is produced that follows the plane. All of these atoms, molecules, and gases are contained within this cloud. Below, in a rectangle with different colors for each exhaust product, you can see how they all stack up against one another. An effect that warms the environment is depicted in red, whereas a cooling effect is shown in blue. The effect these gases have on either warming or cooling the atmosphere is highlighted underneath the cloud.

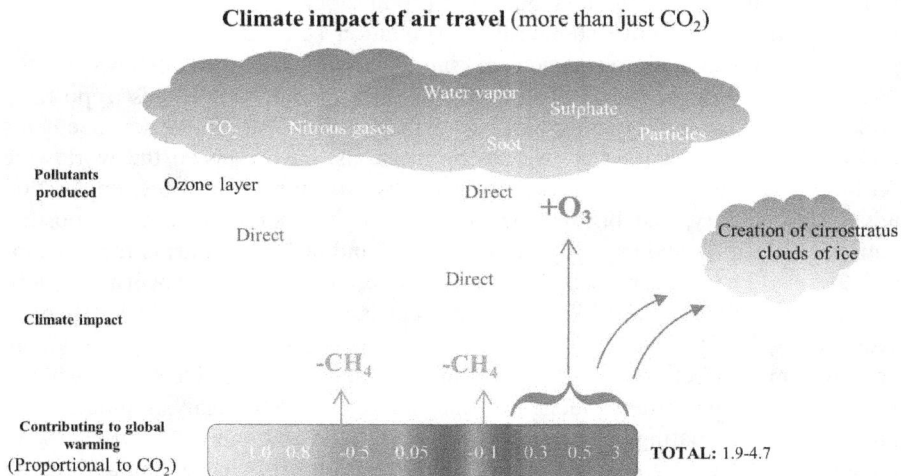

**FIGURE 4.4** A quick look at aviation emissions.

## 4.6    TBC CONTRIBUTING TO REDUCING $CO_2$ EMISSIONS

TBCs may minimize aviation's environmental effects. TBCs based on germinate can meet the requirement for coatings that meet tomorrow's GT engine standards for commercial aircraft. Germinated TBCs can meet market needs due to their decreased thermal conductivity and improved temperature phase stability. The age of the automobiles that will be driven on these engines makes this a necessity. The innovative ceramic material Martin, which is built on germinates, has shown significant advances in its temperature phase stability. When the temperature increases, less power is needed to actively cool GT engines. The engine's $CO_2$ emissions will go down, and its overall performance and efficiency will go up as a result of this change. The new germinate-based TBC was also in line with the $CO_2$ reduction aims of the NASA organization that oversaw the project. Over time, thermionics will be able to mesh with NASA's various emission-free initiatives. More resources are required when more people are living in a given location. The local environment will be altered as a result of changes in the surrounding air and noise quality, as well as the local climate. When calculating the impact of planes that don't spew $CO_2$ into the air, a multiplier technique is sometimes used. Total climate impact caused by airplanes, divided by the aircraft's per-mile $CO_2$ emissions. Pollutants include substances like CO, $NiO_2$ oxides, and C-H. It protects high-temperature-operating components so that they can function at their peak while staying within their operating range.

## 4.7    AI-ML APPROACH FOR SUSTAINABLE GROWTH OF GLOBAL COATING MARKET

In technical sectors like the coatings industry, AI and ML accelerate product development. Design-of-experiment techniques and sophisticated statistical analytics help coating formulators optimize product qualities while satisfying regulatory and sustainability standards. Formulation science uses AI and ML to incorporate more data into decision-making. Polaris Market Research provided the "Artificial Intelligence Market Trend, Growth Drivers, and Challenges" report. By 2026, the worldwide AI market will be worth $54 billion. Robotics advancements and increased use, especially in emerging nations, have boosted the worldwide AI industry. The global AI market is driven by customer experience, application areas, productivity, and big data integration. Governments are now supporting national AI efforts and engaging in a technical and ideological arms race to control the machine learning sector. To lead AI research, the UK government will spend £300 million. The Chinese government is now building a $150 billion AI sector by 2030. However, a scarcity of competent workers and threats to human eminence may affect industry development. Due to sophisticated technology, these issues should have little influence on the market. A PMR analysis predicts the global paint and coatings industry will reach $ 286.54 billion by 2026. Because of numerous advances driven by a strong rebound in global construction and manufacturing, especially in Western Europe, Japan, and North America, architectural paints and coatings are expected to increase in demand in the forecast period of

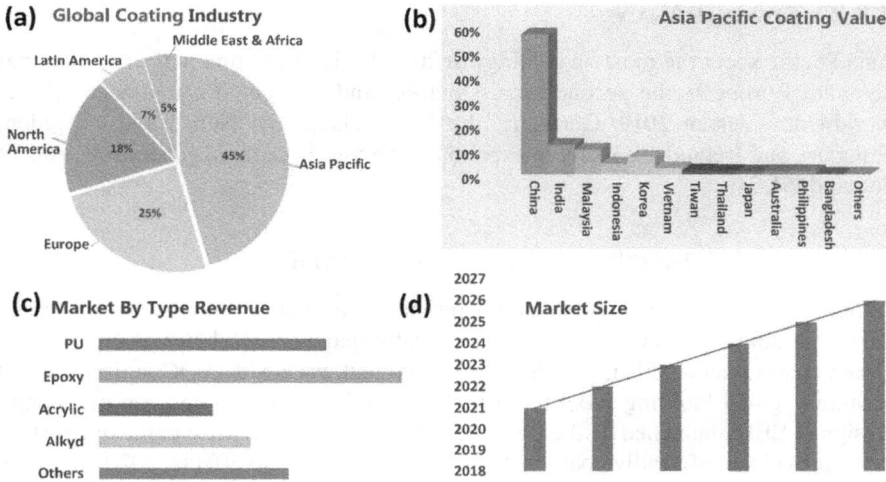

**FIGURE 4.5** Market scenario representing sustainability of thermal spray and coatings in the future [50]. Permissions under the CC BY 4.0 license.

2018–2026. Chemical, automotive, industrial production, etc. will grow faster than their counterparts, which are going to benefit from an improved chance of growing vehicle production each year and industrial activities, including paints and coatings (Figure 4.5) [48,49].

## 4.8 MARKET SIZE AND 2026 FORECAST

Between 2018 and 2026, the worldwide paint and coatings market will rise by 6.0%. North American and Western European building construction advancements will influence the evaluation. The rising global manufacturing base drives demand for coatings used in automotive, durable goods, and industrial maintenance applications. Paints and coatings make other items better, safer, more appealing, durable, and sellable. These items enable manufacturing and add value to goods. Coatings sales are rising as home construction resumes worldwide. However, most advancements occur in the automotive industry, which is crucial for market development throughout the predicted period [49,51].

### 4.8.1 SEGMENTATION

Resin, technology, and end-use separate worldwide coatings. Acrylic, epoxy, polyurethane, alkyd, polyester, and other resins are subdivided [52,53]. Water-based, solvent-based, powder coating, high solids, and other technologies follow. Architectural and nonarchitectural paint and coating end-users exist. Acrylic items dominated consumption and prices. Global solvent growth is predicted to be mild. VOC rules are tightening [54,55].

### 4.8.2   REGIONAL ANALYSIS

Asia Pacific spent the most on coatings in 2017. India and China will drive regional demand. Europe is the second-largest market and accounted for almost 27% of worldwide sales in 2019. Germany, the UK, Poland, the Netherlands, Sweden, Hungary, and Ireland are likely to expand their building sectors, generating product demand [51,54].

## 4.9   GREEN COATING MARKET WORLDWIDE

Green coatings, ecologically friendly paint materials, were created since consumers are demanding them. VOC reduction in decorative paints necessitates green coatings. Green coating demand is rising due to regional and worldwide VOC legislation and voluntary green building programs like Leadership in Energy and Environmental Design (LEED), launched by the US Green Building Council. Most paint and coating firms provide eco-friendly coatings to prevent toxic negative effects. 2020 US green coatings market was $27.1 billion. It has 29.44% of the worldwide market. China, the world's second-largest economy, is predicted to reach $22 billion by 2027 with a 4.1% compound annual growth rate (CAGR). Japan and Canada have 4.3% and 3.4% growth forecasts for 2020–2027, respectively. Germany will gain 3.6% CAGR, and Europe will reach $22 billion by 2027 [56–58].

## 4.10   SUSTAINABILITY IN COATING INDUSTRY AND AI

According to the researchers, AI models successfully identified such irregularities, allowing the naval, hydropower, wind power, and marine industries to use AI to analyze images over time to better understand corrosion and coating breakage trends. This "digital innovation" in artificial intelligence will remotely examine, identify, and quantify shell cracking and other structural flaws, boosting safety and dependability. AI can aid in maritime rescue, relief, and disaster relief [59,60]. Table 4.1 includes the most promising coating industry strategies.

This chapter highlighted AI and ML's importance in surface coating. The coating industry's technological shift comes from AI and ML adoption. AI's key advancement is its widespread use in numerous industries, and service sectors. Organizations may improve resource utilization, inventory costs, and projections by using AI technology. Researchers are using evolutionary algorithms to automatically identify novel polymers and coating formulas based on a set of material attributes. In flow polymerization, autonomous synthesis, and formulation, scientists are developing integration tools to pair novel models with new experimental methods. These technologies will accelerate high-performance material design and discovery. Convolutional neural networks can classify spray process photos and real-time videos with minimum calibration on enormous datasets in future studies. The approach is simple, but huge datasets, kernels, and feature vectors need additional investigation. The literature study provides a structural similarity measure for coating microstructure comparisons. Automating the laborious microstructural coating characterization procedure may expand standardization. Training

**TABLE 4.1**

**Sustainable AI Approaches Used in Coating Industry Challenges [61]**

| S. No. | References | Description | Approach |
|---|---|---|---|
| 1 | [62,63] | Coating optimization | Semi-analytical experimental based approaches |
| 2 | [64,65] | Robot path optimization | Self-organizing maps and geodesic-based model |
| 3 | [66–68] | Process monitoring & control | Artificial neural networks coupled with sensors |
| 4 | [69,70] | Microstructural characterization & defect detection | Convolution & deep neural networks |
| 5 | [71,72] | Conformance | Computer vision structural similarity |
| 6 | [73] | High fidelity | Image analysis image super-resolution |

*Source:* Permissions from Elsevier.
AI, artificial intelligence.

additional models with coating microstructural characteristics will help improve this technique and completely automate coating fault identification using CNNs. Thus, AI and ML constitute the fourth industrial revolution, allowing us to tackle intractable challenges and build a more inventive environment. ML has shown that it can manage macroscopic to microscopic aims, including those in this review, as it is better understood and studied. Thus, ML may be turned into a fast and reliable materials science tool due to its great applicability. ML advances material science research and theory-building. Thus, this study may guide future researchers to optimize surface coating parameters using AI and ML.

## 4.11 CONCLUSIONS AND FUTURE PERSPECTIVE

When it comes to adding or restoring functionality to a solid surface, surface engineering techniques like thermal spraying are the answer. Thermal spraying stands in sharp contrast to energy-intensive processes like melting, casting, extruding, and welding in terms of their contribution to the cause of global warming. As a result of their inclusion on the supply risk register, a small number of metals and alloys used in the thermal spraying process—among them hydrogen—are now considered important raw materials. To maintain the benefits of thermal spraying as a relatively green approach, efforts to address recycling and reuse and discover alternatives to these are very timely. Incorporating digital technology (ML and AI) and thermal spray coatings would further improve the sustainability of the production process by decreasing the use of gas, powder, and electricity. For governments to reward businesses that function with low GHG emissions, they need more information on raw material utilization and waste, which can be gathered through the digitalization of the thermal spray industry.

## REFERENCES

[1] Prashar G, Vasudev H. Structure-property correlation of plasma-sprayed Inconel625-$Al_2O_3$ bimodal composite coatings for high-temperature oxidation protection. *J Thermal Spray Technol.* 2022;31(8):2385–2408.

[2] Prashar G, Vasudev H. Structure-property correlation and high-temperature erosion performance of Inconel625-$Al_2O_3$ plasma-sprayed bimodal composite coatings. *Surface Coat Technol.* 2022;439:128450.

[3] Singh J. Analysis on suitability of HVOF sprayed Ni-20Al, Ni-20Cr and Al-20Ti coatings in coal-ash slurry conditions using artificial neural network model. *Ind Lubr Tribol.* 2019;71(7):972–982.

[4] van Westing E, Savran V, Hofman J, *Materials Innovation Institute M2i Recycling of Metals from Coatings A Desk Study*; 2013; 4:1515–1529.

[5] Viswanathan V, Katiyar NK, Goel G, Matthews A, Goel S. Role of thermal spray in combating climate change. *Emerg Mater.* 2021; 4:1515–1529.

[6] Acemoglu D, Restrepo P. *Artificial Intelligence, Automation, and Work.* NBER Working Paper No. 24196. National Bereau of Economic Research; 2018. DOI 10.3386/w24196

[7] Bolukbasi T, Chang KW, Zou JY, Saligrama V, Kalai AT. Man is to computer programmer as woman is to homemaker? Debiasing word embeddings. In: Lee D, Sugiyama M, Luxburg U, Guyon I, Garnett R. editors, *Advances in Neural Information Processing Systems*; 2016 (30th Annual Conference on Neural Information Processing Systems 2016, 5–10 December 2016, Barcelona, Spain) Curran Associates, Inc. Red Hook, New York, United States. p. 29.

[8] Norouzzadeh MS, Nguyen A, Kosmala M, Swanson A, Palmer MS, Packer C, Clune J. Automatically identifying, counting, and describing wild animals in camera-trap images with deep learning. *Proc Natl Acad Sci.* 2018;115(25):E5716–E5725.

[9] Tegmark M. *Life 3.0: Being Human in the Age of Artificial Intelligence.* London: Random House Audio Publishing Group; 2017.

[10] Sarc R, Curtis A, Kandlbauer L, Khodier K, Lorber KE, Pomberger R. Digitalisation and intelligent robotics in value chain of circular economy oriented waste management-A review. *Waste Manag.* 2019;95:476–492.

[11] Khan MN, Shamim T. Effect of operating parameters on a dual-stage high velocity oxygen fuel thermal spray system. *J Thermal Spray Technol.* 2014;23:910–918.

[12] Ikeuchi D, Vargas-Uscategui A, Wu X, King PC. Neural network modelling of track profile in cold spray additive manufacturing. *Materials.* 2019;12(17):2827.

[13] Gössling S, Humpe A. The global scale, distribution and growth of aviation: Implications for climate change. Global Environ Change. 2020;65:102194.

[14] El Takriti S, Pavlenko N, Searle SJ, Ràhtosdfp A-A. Mitigating international aviation emissions; Risks And Opportunities for Alternative Jet Fuels, In: The International Counsel of Clean Transportation, Washington, United States, White Paper: March 2017. Repéré à https://theicct. org/sites/default/files/publications/Aviation-Alt-Jet-Fuels_ICCT_White-Paper_22032017_vF

[15] Masson-Delmotte V, Zhai P, Pirani A, Connors SL, Péan C, Berger S, et al. Climate change 2021: The physical science basis; In: Contribution of working group I to the sixth assessment report of the intergovernmental panel on climate change, 02 June 2021, p. 2.

[16] Zeng Y, Friess DA, Sarira TV, Siman K, Koh LP. Global potential and limits of mangrove blue carbon for climate change mitigation. *Curr Biol* 2021;31:1737–1743.e3.

[17] Wang Z, Shuai S, Li Z, Yu WJE. A review of energy loss reduction technologies for internal combustion engines to improve brake thermal efficiency. *Energies.* 2021;14:6656.

[18] Gautam SS, Singh R, Vibhuti AS, Sangwan G, Mahanta TK, Gobinath N, et al. Thermal barrier coatings for internal combustion engines: A review. *Mater Today: Proc.* 2022;51:1554–1560.

[19] Bobzin K, Brögelmann T, Kalscheuer C, Yildirim B, Welters M. Correlation of thermal characteristics and microstructure of multilayer electron beam physical vapor deposition thermal barrier coatings. *Thin Solid Films.* 2020;707:138081.

[20] Keyvani A, Mostafavi N, Bahamirian M, Sina H, Rabiezadeh A. Synthesis and phase stability of zirconia-lanthania-ytterbia-yttria nanoparticles; a promising advanced TBC material. *J Asian Ceramic Soc.* 2020;8:336–344.

[21] Motoc AM, Valsan S, Slobozeanu AE, Corban M, Valerini D, Prakasam M, et al. Design, fabrication, and characterization of new materials based on zirconia doped with mixed rare earth oxides: Review and first experimental results. *Metals.* 2020;10:746.

[22] Goral M, Swadźba R, Kubaszek TJS, Technology C. TEM investigations of TGO formation during cyclic oxidation in two-and three-layered thermal barrier coatings produced using LPPS, CVD and PS-PVD methods. *Surface Coat Technol.* 2020;394:125875.

[23] Besisa DH, Ewais EMM. Advances in functionally graded ceramics-processing, sintering properties and applications. 2016;1:32.

[24] Chen H-F, Zhang C, Liu Y-C, Song P, Li W-X, Yang G, et al. Recent progress in thermal/environmental barrier coatings and their corrosion resistance. *Rare Metals.* 2020;39:498–512.

[25] Lashmi P, Ananthapadmanabhan P, Unnikrishnan G, Aruna ST. Present status and future prospects of plasma sprayed multilayered thermal barrier coating systems. *J Eur Ceramic Soc.* 2020;40:2731–2745.

[26] Chen H, Zhang C, Xuan J, Liu B, Yang G, Gao Y, et al. Effect of TGO evolution and element diffusion on the life span of YSZ/Pt-Al and YSZ/NiCrAlY coatings at high temperature. *Ceramics Int.* 2020;46:813–823.

[27] Lv B, Jin X, Cao J, Xu B, Wang Y, Fang D. Advances in numerical modeling of environmental barrier coating systems for gas turbines. *J Eur Ceramic Soc.* 2020;40:3363–3379.

[28] Rauf A, Yu Q, Jin L, Zhou C. Microstructure and thermal properties of nanostructured lanthana-doped yttria-stabilized zirconia thermal barrier coatings by air plasma spraying. *Scr Mater.* 2012;66:109–112.

[29] Park K, Kim K, Kim D, Moon B, Park S, Seok C-S. Failure mechanism of plasma-sprayed thermal barrier coatings under high-temperature isothermal aging conditions. *Ceramics Int.* 2021;47:15883–15900.

[30] Mathanbabu M, Thirumalaikumarasamy D, Thirumal P, Ashokkumar M. Study on thermal, mechanical, microstructural properties and failure analyses of lanthanum zirconate based thermal barrier coatings: A review. *Mater Today: Proc.* 2021;46:7948–7954.

[31] Keyvani A, Saremi M, Sohi MH. Oxidation resistance of YSZ-alumina composites compared to normal YSZ TBC coatings at 1100 C. *J Alloys Compounds.* 2011;509:8370–8377.

[32] Keyvani A, Saremi M, Heydarzadeh Sohi M. Oxidation resistance of the nanostructured YSZ coating on the IN-738 superalloy. *J Ultrafine Grained Nanostructured Mater.* 2014;47:89–96.

[33] Kumar A, Moledina J, Liu Y, Chen K, Patnaik PC. Nano-micro-structured 6%-8% YSZ thermal barrier coatings: A comprehensive review of comparative performance analysis. *Coatings.* 2021;11:1474.

[34] Lima RS. Porous APS YSZ TBC manufactured at high powder feed rate (100 g/min) and deposition efficiency (70%): Microstructure, bond strength and thermal gradients. *J Thermal Spray Technol.* 2022;31:396–414.

[35] Chen S, Zhou X, Song W, Sun J, Zhang H, Jiang J, et al. Mg2SiO4 as a novel thermal barrier coating material for gas turbine applications. *J Eur Ceramic Soc.* 2019; 39:2397–2408.

[36] Chang SY, Oh K-YJC. Contribution of high mechanical fatigue to gas turbine blade lifetime during steady-state operation. *Coatings*. 2019;9:229.

[37] Okada M, Kitazawa R, Takahashi T, Ozeki TJ. Crack growth in thermal barrier coating subjected to thermal cycling under temperature gradient. *Int J Gas Turbine Propuls Power Syst*. 2022;13:7–13.

[38] Lashmi P, Aruna ST. *An Overview of Plasma-Sprayed Thermal Barrier Coating Activities in India*; 2022, pp. 733–753.

[39] Ramesh M, Marimuthu K, Karuppuswamy P, Rajeshkumar L. Microstructure and properties of YSZ-$Al_2O_3$ functional ceramic thermal barrier coatings for military applications. *Bol Soc Esp Ceram Vidr*. 2022;61:641–652.

[40] Chen S, Zhou X, Cao X, Jiang J, Deng L, Dong S, et al. Novel thermal barrier coatings based on Mg2SiO4/8YSZ double-ceramic-layer systems deposited by APS. *J Alloys Compounds*. 2022;908:164442.

[41] Sadri E, Ashrafizadeh F, Eslami A, Jazi HS, Ehsaei H. Thermal shock performance and microstructure of advanced multilayer thermal barrier coatings with Gd2Zr2O7 topcoat. *Surface Coat Technol*. 2022;448:128892.

[42] Kadam NR, Karthikeyan G, Kulkarni DMJJo M. The effect of spray angle on the microstructural and mechanical properties of plasma sprayed8YSZ thermal barrier coatings. *J Micromanuf*. 2022;5:181–192.

[43] Wang J, Chen M, Wang Y, Li B, Yu Y, Tian Y, et al. Preparation and thermo-physical properties of La2AlTaO7 ceramic for thermal barrier coating application. *Mater Chem Phys*. 2022;289:126465.

[44] Pan Y, Liang B, Hong D, Han D, Zhong X, Niu Y, et al. Effect of TiAlCrNb buffer layer on thermal cycling behavior of YSZ/TiAlCrY coatings on γ-TiAl alloys. *Surface Coat Technol*. 2022;431:128000.

[45] Doleker KM, Karaoglanli ACJS, Technology C. Comparison of oxidation behavior of YSZ and Gd2Zr2O7 thermal barrier coatings (TBCs). *Surface Coat Technol*. 2017;318:198–207.

[46] Lance MJ, Unocic KA, Haynes JA, Pint BA. The effect of cycle frequency, $H_2O$ and $CO_2$ on TBC lifetime with NiCoCrAlYHfSi bond coatings. *Surface Coat Technol*. 2014;260:107–112.

[47] Cui Q, Seo S-M, Yoo Y-S, Lu Z, Myoung S-W, Jung Y-G, et al. Thermal durability of thermal barrier coatings with bond coat composition in cyclic thermal exposure. *Surface Coat Technol*. 2015;284:69–74.

[48] Verma J, Gupta A, Kumar D, Steel protection by $SiO_2$/$TiO_2$ core-shell based hybrid nanocoating, *Prog Org Coating*. 2022;163:1–11.

[49] Sally G. The global coatings market in 2020. *Polymer Paint Color J*. 2020. Available from: https://polymerspaintcolourjournal.com/news/the-global-coatings-market-in-2020.

[50] Verma J, Khanna AS. Digital advancements in smart materials design and multifunctional coating manufacturing. *Phys Open*. 2023;14:100133.

[51] Grand View Research, Industrial coatings market size, share & trends analysis report by product. 2020. Available from: https://www.grandviewresearch.com/industry-analysis/industrial-coatings-market.

[52] Verma J, Nigam S, Sinha S, Bhattacharya A, Comparative studies on poly-acrylic based anti-algal coating formulation with SiO2@ TiO2 core-shell nanoparticles. *Asian J Chem*. 2018;3:1120–1124.

[53] Verma J, Nigam S, Sinha S, Sikarwar BS, Bhattacharya A, Irradiation effect of low energy ion on polyurethane nanocoating containing metal oxide nanoparticles. *Radiat Eff Defects Solids*. 2017;172:964–974.

[54] Coating world: Market report; 2021. Available from: https://www.coatingsworld.com/breaking-news/market-reports.

[55] Coating world: Green coating; 2017. Available from: https://www.coatingsworld.com/issues/2017-11-01/view_features/green-coatings.

[56] Data Bridge Market Research, Global green coatings market: Industry trends and forecast to 2028;Report, 2020. Available from: https://www.databridgemarketresearch.com/reports/global-green-coatings-market.

[57] Globe News Wire, Green coatings - Global market trajectory & analytics, ResearchAndMarkets.com's;Report, 2020. Available from: https://www.globenewswire.com/en/newsrelease/2020/11/27/2135264/28124/en/Global-Green-Coatings-Industry-2020-to-2027-Market-Trends-and-Drivers.html.

[58] Market Research Future, Green coating market: Information by type, end-use industry and region; 2021. Available from: https://www.marketresearchfuture.com/reports/green-coating-market.

[59] Donlon M, AI to detect corrosion, coating wear in ships, offshore structures; 2019. Available from: https://insights.globalspec.com/article/11720/researchers-use-ai-to-detect-corrosion-coating-wear-in-ships-offshore-structures.

[60] Verma J, Gupta A, Kumar D, Steel protection by SiO2/TiO2 core-shell based hybrid nanocoating. *Prog Org Coating* 2022;163:1–11.

[61] Konstantina M, Konstantinos D, Bryan A, Spyros K. Digital transformation of thermal and cold spray processes with emphasis on machine learning. *Surf Coat Technol.* 2022;433:128138.

[62] Wu H, Xie X, Liu M, Chen C, Liao H, Zhang Y, Deng S. A new approach to simulate coating thickness in cold spray. *Surface Coat Technol.* 2020;382:125151.

[63] Tzinava M, Delibasis K, Allcock B, Kamnis S. A general-purpose spray coating deposition software simulator. *Surface Coat Technol.* 2020;399:126148.

[64] Tzinava M, Delibasis K, Kamnis S. Self-organizing maps for optimized robotic trajectory planning applied to surface coating. *IFIP Adv Inf Commun Technol.* 2021;627:196–206.

[65] Guessasma S, Montavon G, Coddet C. Neural computation to predict in-flight particle characteristic dependences from processing parameters in the APS process. *J Thermal Spray Technol.* 2004;13:570–585.

[66] Kothuru A, Nooka SP, Liu R. Audio-based tool condition monitoring in milling of the workpiece material with the hardness variation using support vector machines and convolutional neural networks. *J Manuf Sci Eng Trans ASME.* 2018;140:111006.

[67] Kamnis S, Malamousi K, Marrs A, Allcock B, Delibasis K. Aeroacoustics and artificial neural network modeling of airborne acoustic emissions during high kinetic energy thermal spraying. *J Thermal Spray Technol.* 2019;28:946–962.

[68] Chowdhury A, Kautz E, Yener B, Lewis D. Image driven machine learning methods for microstructure recognition. *Comput Mater Sci.* 2016;123:176–187.

[69] De Albuquerque VHC, Cortez PC, De Alexandria AR, Tavares JMRS. A new solution for automatic microstructures analysis from images based on a backpropagation artificial neural network. *Nondestr Test Eval.* 2008;23:273–283.

[70] Wang Z, Bovik AC, Sheikh HR, Simoncelli EP. Image quality assessment: From error visibility to structural similarity. *IEEE Trans Image Process.* 2004;13:600–612.

[71] Avanaki AN. Exact global histogram specification optimized for structural similarity. *Opt Rev.* 2009;16:613–621.

[72] Jung J, Na J, Park HK, Park JM, Kim G, Lee S, Kim HS. Super-resolving material microstructure image via deep learning for microstructure characterization and mechanical behavior analysis. *NPJ Comput Mater* 2021;7:96.

[73] Dong C, Loy CC, He K, Tang X, Image super-resolution using deep convolutional networks. *IEEE Trans Pattern Anal Mach Intell.* 2016;38:295–307.

# 5 Role of Machine Learning Techniques in Coating Process Monitoring, Controlling, and Optimization

## ABBREVIATIONS

| | |
|---|---|
| **AI** | Artificial intelligence |
| **ATI** | Aircraft Tooling Inc. |
| **Cobots** | Collaborative robots |
| **HRC** | Human–robot collaboration |
| **HVOF** | High-velocity oxy-fuel |
| **LfD** | Learning from demonstration |
| **ML** | Machine learning |
| **PS** | Plasma spray |
| **SOD** | Stand-off distance |

## 5.1 INTRODUCTION TO ROBOTS

Nowadays, automation is used in almost every industry to improve the efficiency, effectiveness, and precision of products and services [1]. Since cyber-physical systems and the Internet of Things have been used in manufacturing and automation systems, the industrial sector has been on the rise [2]. Self-contained flows is a model for physically autonomous processes with stringent network capabilities and user-friendly interfaces. When characterized according to natural human–machine interfaces, i.e., those that lessen the technical barriers necessary for engagement, the interactive component of self-contained flows achieves its zenith.

Cobots, or collaborative robots, are industrial systems built for use in robot-human contexts [3]. ISO/TC 299 describes cobot as a robot built to collaborate with people [4]. Cobots are universally adapted to replace humans in various processes in which high precision is required. Cobots and robots are extensively used in harsh environments in the industry [5]. Thermal spraying, cutting, machining, and welding are examples where the risk of an accident is high. Robots are being frequently adopted in industry after the commencement of the twenty-first century. However, robots are extensively used in thermal spraying for the uniformity and durability of coating

DOI: 10.1201/9781003400660-5

layers. In this environment, the robot works with human operators, whose efforts are augmented by the cobot's unique capabilities. Cobots have found their use in several fields, including the car industries, surgery, educational tools, the coating and welding industry, assembling in many industries, and rehabilitation [6]. Cobots are designed to adapt to the shifting requirements of mass customization, a trend that the business world is eager to capitalize on but that is beyond the purview of industrial robots designed for mass production. Companies are adopting them because they can replace humans in repetitive or poorly ergonomic tasks, they can share assembly lines with humans in confined spaces, they require fewer safety precautions than simple robots, they can quickly adapt to changing workloads, and they can be programmed with minimal human input. The Internet of Things, smart manufacturing, and cloud-based manufacturing are all examples of the real-world advancements that make up the theoretical notion of Industry 4.0. To achieve continuous improvement, prioritize value-added operations, and minimize waste, the term "Industry 4.0" describes the practice of tightly integrating humans into the production process [7]. In the context of fast industrial growth, collaborative robots in industrial automation have evolved. Industry 4.0 presents the innovative idea of people and robots working side-by-side in the same physical environment. Due to the intimate partnership between people and machines, research in the area of industrial robotics often focuses on the development of safe human–machine interaction systems. When considering solutions to difficulties in industrial robotics, it's important to consider both technological and societal factors. The introduction of robotic solutions, such as those for production optimization and automation, is also on the list, and it is anticipated that cobots will form the backbone of a significant commercial expansion shortly [8].

As mass bespoke production becomes the norm, small- and medium-sized enterprises have embraced agile manufacturing practices, leading to the rise in the popularity of collaborative robots in the workplace. However, there is an absence of highly qualified staff to program the robot, do the complicated tasks, and establish the connection of robotic systems to other smart devices in the factory. These are the key hurdles in the industrial adoption of cobots. Since many collaborative robot systems are meant to be programmed by professionals rather than regular employees, teaching and simulating them by non-robotics experts is a significant difficulty [9]. The primary objective of human–robot collaboration (HRC) is to facilitate human–robot interaction in a risk-free setting [10]. The manufacturing process does not go from being entirely human-driven to being entirely automated without any human interaction whatsoever. Because of safety regulations, this process comes with a few drawbacks. For instance, when the operating team is not present, the machine may run automatically. In collaborative robotics, people and machines work side-by-side in the same environment so that the robot may assist with tasks that are not ergonomically optimal, tedious, unpleasant, or even hazardous. Through the use of cutting-edge sensors, the robot keeps track of its motions to ensure that it poses no limitation to, and most importantly, no risk to, the human worker [11].

In the manufacturing industry, cobots are highly sought after because of their lower cost, built-in safety features, and user-friendly interfaces. Even more so for small- and medium-sized enterprises, who often struggle to automate their production using conventional robots [12]. Companies in the mass production sector and the automobile industry in particular are keen to adopt HRC as a means of increasing their competitiveness and ushering in the next phase of automation and industrial development, known as Industry 4.0. When it comes to installing sound and moisture insulation on the interior of automobile doors, for example, the BMW Group's Spartanburg factory used cobots to enhance ergonomics [13]. At the Audi Brussels factory, MRK-Systeme KR SI cobot is used to glue the different reinforcement plates [14]. Typical MRK-Systeme KR SI cobots have been shown in Figure 5.1a. The Fanuc CR-35iA cobot is commonly used in many industries for handling duties to resolve ergonomic challenges. In the automotive industry, the Fanuc CR-35iA cobot is used in the assembly line for lifting spare tires into vehicles [15]. Typical Fanuc CR-35iA cobots have been shown in Figure 5.1b. At the Volkswagen factory in Wolfsburg, Germany, a KUKA cobot is used to conduct screwing on a drive train in hard-to-reach places [16]. In addition, Skoda has used the KUKA cobot to assist human workers in the assembly of direct-shift gears [17], as shown in Figure 5.1c. ABB YuMi is used for complex assembly applications in the industry [18]. A typical ABB YuMi cobot is shown in Figure 5.1d. Audi's UR3 cobot, designed to spread glue on vehicle roofs, eliminates all kinds of physical barriers [19]. The human works close to the cobot, while the latter screws in more inconvenient places. The UR10 cobots were installed at Nissan's massive Yokohama facility to loosen bolts and transport heavy components, freeing up workers to focus on other, more productive duties [20]. A typical UR series cobot is shown in Figure 5.1e. Safety when working alongside

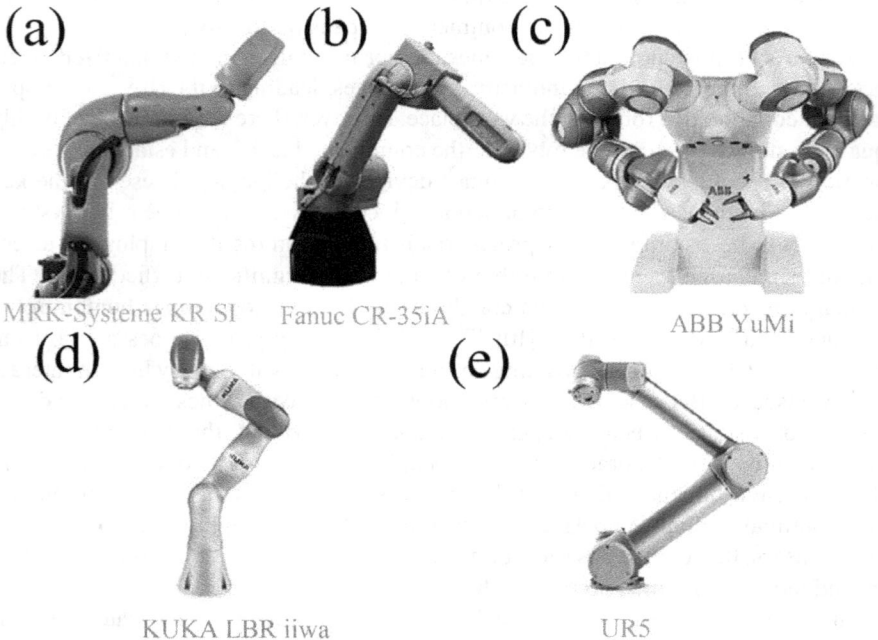

FIGURE 5.1   Different types of cobots used in various industries [24]. Permissions from MM Science.

people is an essential need for the aforementioned applications of cobots [21]. They may not necessarily need high levels of intelligence, human-like awareness, or the ability to make sound decisions, however. Since the pieces to be handled are maintained at known locations, the human and cobot may perform their duties independently, and the cobot can stick to a pretty set action/motion plan [22]. Thus, it is evident that "Independent" or "Simultaneous" implementations of HRC situations are more common in industry [23]. However, most such implementations fall short of showcasing the usefulness and adaptability of cobots in a largely unstructured work environment since they impose extra limits on the cobots' surroundings (in terms of fixed pieces or equipment placements).

## 5.2 ROBOTICS IN CONTROLLING THE COATING PARAMETERS

When applying thermal or cold spray coatings to the surfaces of complicated objects, it is vital to make use of offline path planning and industrial robots to achieve a uniform thickness throughout the coating. When this occurs, the kinematic and dynamic performance of industrial robots might be adversely affected to a large degree. For example, it results in a large number of unintended reorientations of the robot's axis, which modifies important handling parameters and may affect the quality of the coating [25]. Figure 5.2 is a schematic representation of the primary handling characteristics that have the potential to alter the coatings.

### 5.2.1 SPRAY ANGLE

The spray angle is the angle in degrees that exists between the axis of the spraying cannon and the substrate surface, as seen in Figure 5.2. When spray particles hit the component surface and transmit their momentum and heat energy to the component surface, the angle should measure 90° in an ideal situation. Because of the restricted mobility of a robot system and the impossibility of maintaining a spray angle of 90°

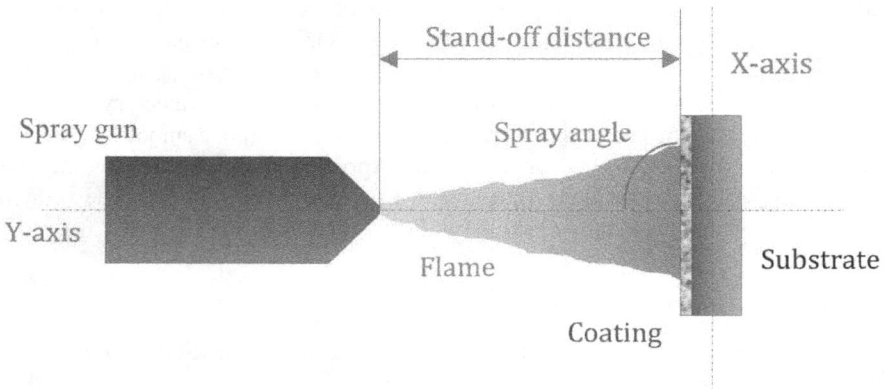

**FIGURE 5.2** Schematic diagram of major handling parameters in thermal spraying.

consistently, the coating of complexly shaped geometries requires that the spray angle be decreased. This shift in impact angle has the potential to bring about unwelcome differences in the deposition rate or the characteristics of the final coating. It had previously been shown via the work of a variety of researchers [26–28] that decreasing the spray angle results in an increase in coating porosity. This happens because the impacting particles lack the necessary energy and the normal velocity component becomes lower, making them unable to adequately cover the whole surface of the substrate, which results in the spraying of coatings with a higher porosity [27]. It is possible to assume that decreased spray angles in TS will affect the efficiency of deposition as well as the microstructure of the coating [29]. Binder et al. [30] observed that deviations from normal impact conditions could significantly alter the deformation behavior of the particle, which resulted in increased porosity levels, and reduced tensile and adhesive strength of cold spray deposits. This was because the particle's deformation behavior was significantly altered (Figure 5.3). Additionally, they reported that when deviations from normal impacts were less than 20° (Figure 5.3c and d), spray particles showed the same deformation behavior as the normal incidence. There were no significant alterations in the levels of porosity, which are required for the majority of the applications. Figure 5.3f illustrates that a tangential component of particle momentum during oblique hits may cause tensile force at the substrate interface. This force is sufficient to dislodge the particles from the surface of the substrate when the impact occurs at an angle of 45°. According to Tillmann et al. [31] and Mostaghimi and Chandra [32], changes in deposition efficiency and coating mechanical properties like micro-hardness, surface roughness, or porosity levels, which can be vital for the final coating quality, occur at spraying angles of 50°. Tillmann et al. [33] concluded that HVOF spraying of fine WC-12Co powders (size 2–10 μm) is less likely to experience variations in the spraying angles as compared to other techniques like plasma or arc spraying. It has been found that there is no substantial loss in coating qualities up to 30°, even though lower spraying angles lead to a fall in deposition rates. On the other hand, the formation of pores and fractures has a detrimental impact on the strength of the coating when the spraying angle is reduced by more than 30°.

From the above literature, it can be concluded that the spraying angle ought to be adjusted in a manner that is normal to the surface of the substrate while the operation is being carried out to produce excellent coatings. On the other hand, carrying this out successfully in the context of intricate geometries is a challenging task. Under these circumstances, there may be some points in the robot's trajectory where the spraying angle can be provided with a slight reduction to obtain a smoother trajectory that is easily reachable by the robot while it is spraying. This would allow the robot to perform spraying operations with greater ease. It enhances not just the spraying process as a whole but also the overall quality of the coating.

## 5.2.2 Stand-Off Distance

During the spraying process, optimizing the stand-off distance (SOD) is one method that may be used to enhance the quality of the coating after it has been sprayed. As shown in Figure 5.2, the SOD refers to the physical distance that separates the spray gun from the substrate that is going to be coated. The SOD has a major impact on the

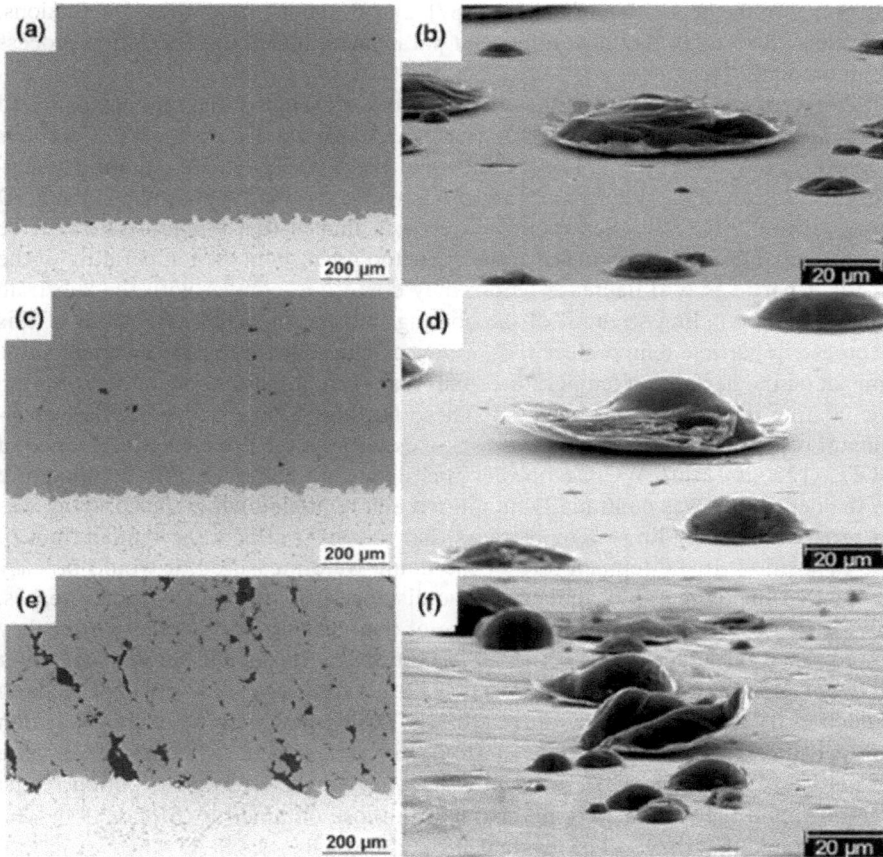

**FIGURE 5.3**   Microstructures of cold spray Ti on AlMg$_3$ specimens for various spray angles; (a and b) normal angle, (c and d) 70°, and (e and f) 45° [30]. Permissions from Springer Nature.

functionality and characteristics of the coatings after they have been sprayed [34]. When SODs are shortened, the resulting temperatures are greater, which in turn leads to the creation of coatings that are more thick and hard [35]. Computational fluid dynamics was used to create models of the as-sprayed HVOF WC-12Co coating, and the results showed that SOD had a major impact on the deposition process [36]. When spraying WC-Co powder, having a large SOD between the gun and the powder leads to reduced velocities, which in turn leads to increased porosity levels. As the SOD increases and the particles in flight grow hotter as they approach the target, the critical velocity that is required for coating buildup and particle deformation gets lower and lower. This results in less particle compaction and decreased contact between splats, which leads to increased levels of porosity. On the other hand, longer stand-off lengths lead to the re-solidification of partly melted particles along the route and create poor deposition efficiencies; hence, they should be avoided if at all possible. Porosity levels are increased when there is insufficient efficiency in the deposition of material [37]. Due to the nature of these interactions, it is clear that one of the most important parameters during the deposition process is an optimal SOD.

To get an understanding of the optimal stand-off lengths that may be exploited to improve the performance of coatings, research was carried out on WC-Co alloys and alloys based on nickel. The producers of the Sulzer Met. Co. equipment suggest keeping an SOD of between 9 and 12 in. while spraying Alloy-625. If the SOD is significantly shorter, then there is a greater chance that the substrate will be overheated. On the other hand, if the SOD is too great, then the temperature of the in-flight particles will decrease before they come into contact with the substrate surface. The bonding strength of the coating will be impacted as a result of this decrease in particle temperature [38]. SOD has been shown to have a considerable impact on the deposition temperature, and as a result, it plays a role in determining the coating quality that is deposited. These findings were discovered via experimental research that was published by Stokes [39]. At the predetermined range of SODs (125–260 mm), WC-Co powder was deposited using TS. The temperature of the deposition was continually monitored and regulated while the spraying was in progress. The findings demonstrated that the SOD affects the temperature of the deposition. If the temperature of the flame were measured, it would be easy to see that the temperature reduces as the distance from the gun's head increases. As a direct result of this, the temperature of the particles will decrease after they have passed through the spray nozzle. According to Stokes and Looney, a greater SOD enables the in-fight sprayed particles to quickly cool down, which leads to a decline in deposition temperature [40,41]. This is the outcome of the reduction in deposition temperature. Yilbas et al. [42] investigated the effects of SOD on a nickel-based alloy. The difference in SOD affected the stress distribution of the coatings, and the results were comparable to those obtained by Stokes [39], who demonstrated that the level of residual stress drops when the SOD ranges from 180 to 200 mm. Therefore, the optimal spraying distance should be maintained during the deposition to provide coatings of higher quality.

### 5.2.3 SPRAYING ROUTE

As can be seen in Figure 5.4, the spraying route for the coating process is often configured in the form of a meander. It is accomplished by making a series of parallel horizontal passes with the spray torch over the surface of the substrate. The time that elapses between each of the two successive passes is referred to as the scanning step. In addition to this, the scanning step that is performed has an effect not only on the porosity but also on the thickness of the coating [43]. The effective management of residual stresses in the coatings after they have been sprayed is an essential component in the production of high-quality coatings. The adhesive strength of final coatings, as well as their resilience to thermal shock, fatigue, and wear, are all severely impacted by residual stresses, along with several kinds of failures, such as buckling deformation, splat de-bonding, and even cracking [44–46]. Regarding the developed residual stresses, several research publications have brought attention to the role that the spray path plays. Tian et al. [47] looked into the effect of the spray path on the

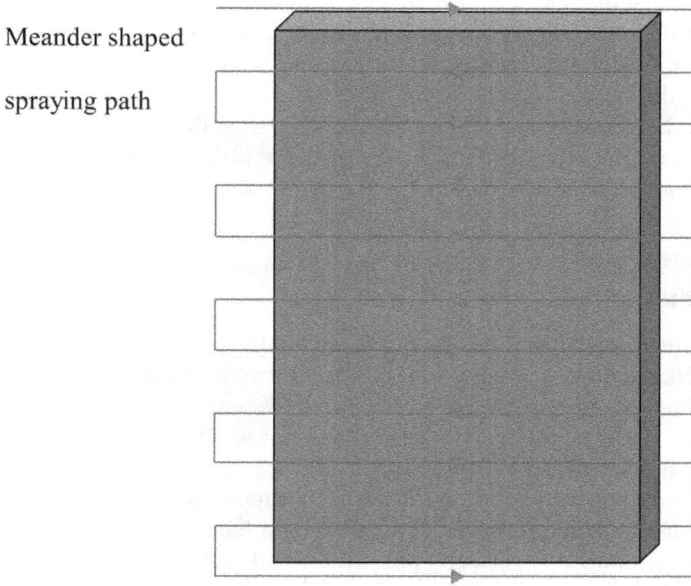

Meander shaped

spraying path

**FIGURE 5.4**  Spraying route for the coating process is often configured in the form of a meander.

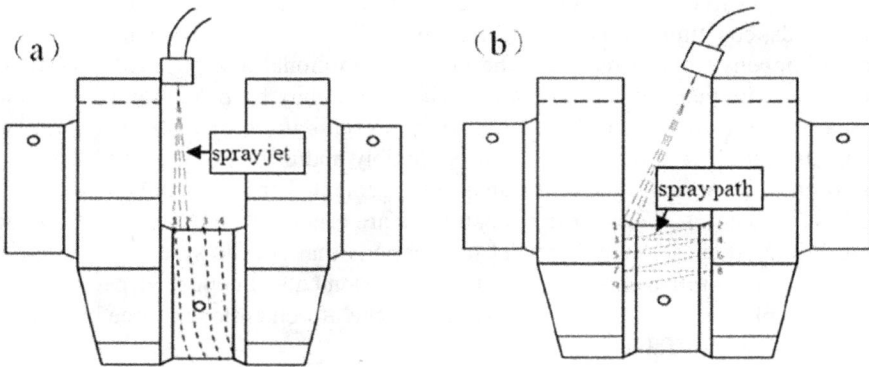

(a)

spray jet

(b)

spray path

**FIGURE 5.5**  Diagrammatic set-up of the spray paths: (a) Circular, and (b) Z-shaped [47]. Permissions from Elsevier.

distribution of residual stress in an electric arc sprayed coating on the surface of the crankshaft journal. Because of the intricate geometry of the crankshaft, offline programming was chosen rather than its online counterpart [48]. A path in the shape of a Z and a path in the form of a circle were chosen to be the spray paths, as shown in Figure 5.5.

The authors concluded that the deposition temperature was relatively lower when the spray route during the spraying process was curved like a "Z," which resulted in the creation of small residual stresses in the coating that was coated. The effect of the spray path employed for the deposition of the coating process depends on both the temperature gradient and temperature level in the coating, as well as the relationship between these factors and the generation of residual stresses in the coating [48]. These investigations shed light on the need for careful design of the spray route in TS and the role that it plays in determining the quality of the coating produced.

## 5.3  ROBOT AND COBOT PROGRAMMING FOR INDUSTRIAL APPLICATIONS

Programming a cobot includes giving it the intelligence to perceive its surroundings and take actions that get the system closer to its intended collaborative purpose. In the past, industrial robot programs have only required the participation of a human (the programmer) in the offline phase. Unless a mistake arises and debugging is required, these programs cannot be modified, while they are being executed, making them rigid and lacking in human awareness. Consequently, a robot operates in a deterministic setting where a human operator does not play a role. The environment is perfectly predictable and stable except in HRC, where an operator introduces randomness and uncertainty. When it comes to programming robots, humans aren't only playing the offline role they usually do with other robots. At runtime or online, the operator may modify the cobot's code. The programming of the cobot is vulnerable to both direct and indirect interference by an operator. The human's direct connection with the cobot, in the form of either information or commands, constitutes this kind of overt participation. A cobot's implicit participation happens when it monitors the human's emotional and physical conditions and adjusts its behavior appropriately. The policy may be programmed by hand or learned from historical data. This study presents three distinct programming elements that allow the cobot to function flexibly and/or be programmed logically based on this varying degree of human engagement. For sequential and supportive HRC situations, these coding capabilities are crucial. Cobots are programmed to recognize their surroundings and take actions that get the system closer to its intended collaborative end state. The following elements of code have been singled out: Examples of cobot programs that feature the aforementioned three characteristics are shown in Table 5.1.

### 5.3.1  COMMUNICATION

The cobot is managed by a human user through some kind of communication channel, either verbal (speaking) or nonverbal (gestures, eye contact, etc.). Gestures, eye contact, body language, head orientation, tactile interfaces, and virtual reality are all examples of nonverbal communication. The programmer works offline to design probable cobot behaviors and fundamental motion control. When operating online, the operator often takes an active role by explicitly commanding the cobot to carry out certain tasks. The programmer's offline job is to specify the actions available to cobot and the motion control it will use.

**TABLE 5.1**

**Some Types of Cobot Programs**

| S. No. | Feature | Offline Programmer's Role | On-Lin Operator's Role |
|---|---|---|---|
| 1 | Communication [49] | Develop an algorithm for voice recognition and an action plan | Execute the job plan via the use of vocal orders (explicit participation) |
| 2 | Optimization [50] | To have the cobot choose the best item to grab, you must create a cost function and an optimization algorithm | By picking up things, you may indirectly influence the cobot's choice of what to pick up by changing its cost function |
| 3 | Learning [51] | Develop a proof-of-concept for Learning from Demonstration (LfD), and show how it works | Give the cobot instructions that it may use to change the order of its actions during construction (direct participation) |
| 4 | Learning [52] | Develop a proof-of-concept for LfD, and show how it works. Do what demonstrators would have you do | Cobots are designed to work alongside humans by watching their actions and mimicking them (implicit participation). The cobot chooses the optimal object to grasp |

### 5.3.2 OPTIMIZATION

A cobot's environment, including barriers and the location of its tools, is mathematically modeled as a result of the cobot's motions. Those come together to make up the cost functions that are optimized to provide the desired results. The cobot's code may be adjusted to lessen the toll on the human operator in terms of effort, energy, and time spent, or to increase the worker's sense of safety and confidence in the system, the quality of output, etc. The developer creates cost functions and optimization algorithms during the offline phase of the project. An operator's presence in a cost function during operation has a significant, though often unnoticed, effect on a cobot's efficiency. The approach has the potential benefit of being more efficient than a human operator.

### 5.3.3 LEARNING

A cobot learns a new task in the same ways a person would, such as through observation, practice, mistakes, correction, and inquiry. A programmer's offline responsibilities include formulating the learning algorithm and supplying the seed data from which the cobot will learn. For example, you may provide examples, go through some trial and error to arrive at a policy, provide some training data, etc. By giving extra information, such as feedback, responses to inquiries, individualized demonstrations, etc., an operator may be able to overtly change the cobot's policy during execution. In addition, the operator may play a role as a prior in the probabilistic learning process of the cobot, influencing it in a roundabout way just by being there in the observable environment.

## 5.4  OFFLINE PROGRAMMING FOR COATING PROCESS

In general, the spraying parameters used during the spraying process for the development of high-quality coatings are selected by the method of "trials and errors" involving the numerous process control methods. This is done to ensure that the coatings produced are of the highest possible quality. Nevertheless, these tests are often carried out in production booths with the assistance of an industrial robot. This strategy is not only incredibly expensive but also very time-consuming in the majority of situations. Therefore, to ease the manufacture of the coating, a variety of models and software tools are employed to simulate the whole process of deposition and forecast the exact coating attributes that are needed. Offline programming is one of these sophisticated methods, and it is the one that will provide a comprehensive solution for the TS process. This solution will cover everything from the production of the spray route to the parameters for simulation to the optimization of the trajectory. It is always possible to build the robot's trajectory for the deposition of coating with the assistance of the actual geometrical model of the component [1], which ensures both the correctness and precision of the route. In Figure 5.6, a diagrammatic representation of this full procedure may be found in the form of an illustration.

### 5.4.1  COMPUTER AIDED DESIGN (CAD) FILE ACQUISITION

To design the trajectories for robots, graphical programming has to have access to the CAD geometry of the substrate. Consequently, the first step is to get the geometric model in 3D. If the original CAD model cannot be located, a simplified version will need to be crafted using CAD software such as Catia, SolidWorks, and Pro-E, among others. If the component cannot be developed using CAD software due to its level of complexity, an alternative method known as reverse engineering will be used. Either a coordinate measuring machine or a laser scanning system may be used to gather the necessary geometrical information about the component [53]. Both of these systems have their advantages. These measured locations may be used to construct the three-dimensional geometric model, which is a more effective way of dealing with complicated components.

### 5.4.2  SELECTION OF OPERATING PARAMETERS FOR THERMAL SPRAYING

According to what is stated in the relevant literature, the parameters for the TS process may be broken down into a few distinct categories: energy parameters, injection parameters for feedstock powder, and other kinematic parameters. The parameters that were discussed earlier can be controlled in one of two ways: either (a) directly, such as the speed of the torch, the spraying distance, and the scanning step, or (b) indirectly, such as the speed and temperature of the particles in flight, among other things. The performance of the procedure as well as the qualities of the coating may be impacted by all of these handling aspects. An optimization approach was suggested by Kout and Müller [54] as a means of calculating and approximately estimating the necessary coating thicknesses for the respective operational parameters. The authors researched the methods of designing path-spray orientation coatings. In a different investigation, Trifa et al. [55] studied an interaction between the operating parameters

- Reverse Engineering
- CAD files

- Obtain workpiece's form

- Select operating parameters
- Analyse workpiece
- Generate curves on surface
- Create points on the curves

- Generate robot path

- Process simulation

- Robot program

- Calibration

- Download robot controller

- Test and application

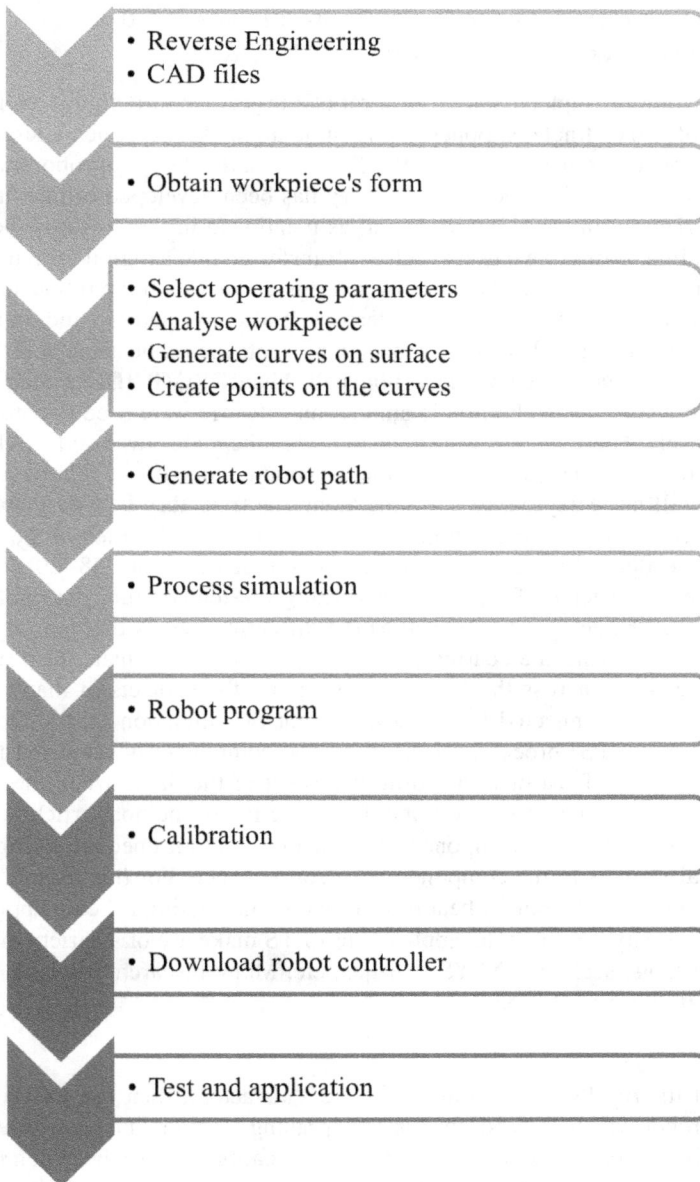

**FIGURE 5.6** Method for the generation of an offline trajectory.

and the characteristics of the deposit. This enables them to select the appropriate settings for the apparatus. Guessasma et al. [56] presented their work, in which they developed an intelligent system based on fuzzy logic to assist in the selection of operating parameters based on the required features and the desired deposition [57]. This system was created to help choose the appropriate values for the parameters.

### 5.4.3  TRADITIONAL MODE OF ROBOT TRAJECTORY IN A VIRTUAL ENVIRONMENT OR SIMULATION

The simulation or implementation of a robot's trajectory in a digital environment consisted of two distinct components: (a) a simulation of robot kinematics, and (b) a simulation of heat transport. During the TS process, the design of robot trajectories plays an important role. Once the trajectory has been developed offline, its execution is carried out in a real environment, as mentioned in the previous section. To begin, the dimensions of a physical cell are transferred into a digital format and used to construct a virtual cell. The working settings under which the robots are placed are recreated in this cell's virtual counterpart. It offers versatility, and, in addition, by maximizing the mobility of the robots, it prevents the robots from colliding with the components that make up the working cell. The ROBOGUIDE (Fanuc M710iC, installed in IFKB) and RobotStudio simulation software were used to construct the virtual models. These models were used to design the real spray booths (ABB Ltd.). However, if any of the axes approach the maximum value of the robotic arm joint, the robot will not carry out any actions along the trajectory beyond that point. If this is the case, then the orientations of some points or the location of the substrate need to be readjusted for that robot to be able to reach all spots [58–60]. The modeling of heat transfer in TS involves analyzing the temperature distribution over the surface of the component as well as the different stresses that are imposed on the component during the coating process. The creation of splats, heat transport, and stress analysis across the interface for TS are the subjects of many kinds of research that are connected to the development of simulation in TS. During the plasma spraying (PS) process, for example, the component is subjected to a heat load for a variety of reasons, including the effect of the flame when the temperature is raised to an increased level and the projection of the hot particles. It brings about structural changes, component deformation, and an unequal distribution of the residual stresses in the component as a consequence. For this reason, simulating and controlling the rate of heat transfer while the coating is being applied is an absolute necessity. Commercial applications of TS make use of a variety of thermal simulation tools, such as ANSYS, Abaqus, etc. Many research articles have been published in this area [61–65].

- Calibration

    If the results of the simulations are satisfactory, then the software for the robot will be placed on it at the spraying location. The software cannot be run on the actual robot until offline calibration has been completed first. It is a necessary step that must be completed before the exam and the application.
- Procedures for Inspection and Application

    Following the tool center point calibration and the component location, the testing of the robot program at a slow speed begins, and once completed, the robot is ultimately prepared for the real spraying operation. However, if the results of the simulation are not satisfactory, it is necessary to return to the phases that came before it to alter and check the robot program.

## 5.5 IMPORTANCE OF PROGRAMMING FOR ROBOTS

This section addresses the research that has been conducted about the significance of offline programming in the field of TS technology. The necessity for robot trajectory optimization to obtain a uniform coating thickness and the following investigation are connected to the development of trajectories in irregularly rotating components via the use of offline programming. In recent years, there has been significant growth in the use of high-accuracy robots in TS applications. When a robot system is utilized for TS, one of the most important parts of the process is the development of the trajectory that the robot will follow. The torch speed, spraying distance, and spraying angle, as well as a few other factors, are among the most important operational parameters involved. The speed at which the torch is moving is the most critical operational parameter to consider while carrying out the TS procedure. When applying excellent coatings, the movement of the torch should be constant, and the direction in which the spraying is done should be as close to normal as possible to the surface of the coating. On the other hand, when the spraying torch is used on a curved component and follows a trajectory, there is a significant amount of change in the orientation of the torch, which results in an apparent reduction in the torch speed. This is seen in Figure 5.7. Because of this, optimizing the movement of the robot is very necessary to get a uniform thickness of the coating [58]. The cobot path can be controlled by using the TP-learning from demonstration optimization algorithm [66]. To improve the computing efficiency of the cobot's route creation, this reinforcement learning is aimed at eliminating extraneous task characteristics found in demos.

## 5.6 SCOPE OF COBOTS IN THE COATING INDUSTRY

Coating and painting cobots are most often used to apply coatings and/or paints to different types of machinery and components. Painting cobots are widely used in the automobile and aircraft industries. Polymeric coatings are applied to automobiles for heavy-duty applications such as finishing look, corrosion protection, and water

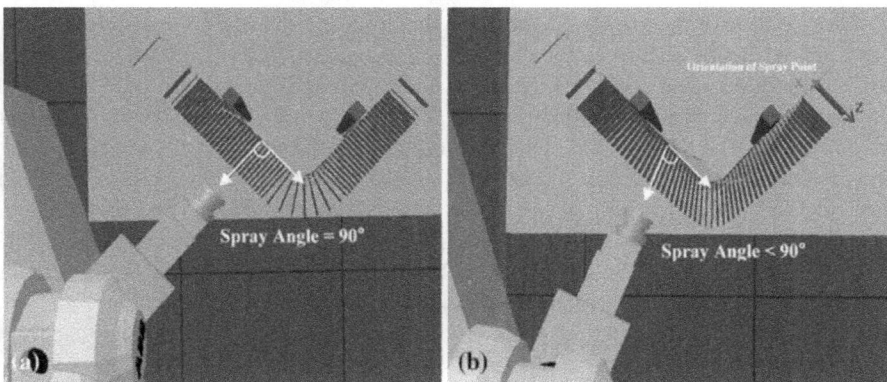

**FIGURE 5.7** (a) The real component with the bigger angle, and (b) the trajectory with the greater angle [58]. Permissions from Springer Nature.

resistance. Aircraft coatings, however, serve several functions, including protection from corrosion and erosion, camouflage, radar attenuation, and even corporate branding. Since an Airbus 380, for instance, has a painted area of around $4400\,m^2$, and a typical aircraft is stripped and repainted every 5 or 6 years, applying and removing paint are both significant activities in the aerospace industry [67]. There are a lot of metal coatings, ceramics, sealants, and paints used in the final stages of airplane manufacturing. When the airplane structure is completed, it becomes extremely difficult for workers to reach 40 ft high and apply these paints and sealants in hard-to-reach areas. Although they can also be used to put on a variety of other coatings such as anti-fingerprint, anti-fog (for glass), water-resistant, sound-absorbing, vibration damping, anti-bacterial, thermal barrier, wear-resistant, etc. Boeing has installed a robot painting system from ABB as part of its attempts to speed up the manufacturing of the 777X. Coating an airplane is an extremely time-consuming procedure that requires numerous coats, up to six times. Two small, light-weight ABB IRB 5500 robots were used to apply Boeing's automated spray system for painting the 32.3-m-long 777X wings in a single position, which resulted in a 75% decrease in floor area and a 100% improvement in throughput [67]. The initial coat application by hand takes 4.5 hours, whereas the robots only need 24 minutes. The robotic system washes the wings, applies a solvent, rinses them, and then paints them twice. However, certain manual preparations, like masking, are still required. Encore Automation, a systems integrator, has also created a commercial aircraft painting robot. The first system was deployed at Boeing's Salt Lake facility and has two FANUC P-250iB robots on 40-m rails with a 3-m lift.

Aircraft Tooling Inc., an aircraft maintenance facility in Dallas, was startled to discover that Universal Robots could handle the high temperatures and harsh atmosphere required for metal powder and PS procedures. After three years of use, the UR "cobots" haven't needed repairs or maintenance. It was just too spotless to be real. When Juan Puente first learned about the Universal Robots robot arms, it was his first reaction. Puente works at Aircraft Tooling Inc. (ATI) in Dallas as the supervisor of the TS department [68]. Figure 5.8a shows the TS process at Aircraft Tooling Inc. is supervised by Juan Puente, who after turning on the HVOF flame, exits the booth to let the UR10 robot do the spraying. New HVOF (high-velocity oxygen fuel) and PS on components were being considered for automation by the firm to meet aviation industry standards for repair [68]. A picture of HVOF and PS using the UR10 cobot is illustrated in Figure 5.8b and c. The ones we looked at made of cast iron were too expensive, too cumbersome to carry from cell to cell, too difficult to program, and too safety-guarded to function in our rather compact spray cells. The UR10 cobot was half the price of the competition and had the necessary reach for the spray distance. It was light and convenient to carry along. The UR robots fall under the category of "collaborative" because of the safety feature that causes the robot arm to cease functioning if it comes into contact with an item or person. ATI was first skeptical about whether or not it would function successfully in the spray booth's severe heat and dust [68]. The robot seemed like it would break under the pressure. Tungsten carbide, a very hard metal, is used in several of these powder coatings. It might leak into the robot's bearings and ruin them. ATI removed the UR10's seals and discovered the bearings to be in good working

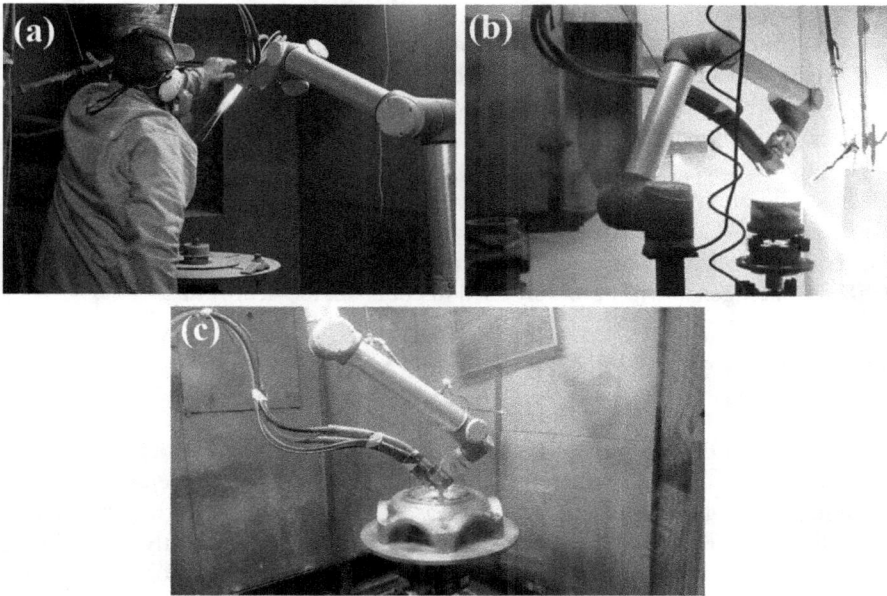

**FIGURE 5.8**  (a) TS process at Aircraft Tooling Inc., (b) HVOF spray using cobot, and (c) PS using UR10 cobot. Open Access Domain by CTE MAG.

order. Adding the recoil from the spray cannon was another issue of worry. It was concerned that the robot's servos may malfunction due to the recoil.

## 5.7  BENEFITS OF AI IN COBOTS

In recent years, the use of AI has surged in various industries, including the manufacturing sector. As a result, the technology is now being implemented in collaborative robots (cobots) to help businesses increase productivity and reduce costs. The following topics can be considered as benefits of implementing AI in cobots:

- Greater precision and accuracy are the primary advantages of incorporating AI into cobots. AI allows cobots to monitor their surroundings for any changes and then react with pinpoint accuracy. This makes the robot more accurate and precise than conventional robots, which reduces the likelihood of mistakes and boosts productivity.
- Enhanced security is another perk of using AI in cobots. Cobots' AI enables them to recognize threats and respond appropriately. This safeguards both human employees and the machinery it operates on by preventing the cobot from causing harm.
- Better scalability is the third advantage of incorporating AI into cobots. Cobots may be developed with AI to do increasingly sophisticated tasks and take on additional labor. In this way, firms may expand their operations without spending more money on machinery or personnel.

- Lastly, enhanced decision-making is the fourth advantage of incorporating AI into cobots. Cobots' decision-making capabilities are greatly enhanced by their access to artificial intelligence. The improved speed and precision with which organizations can now make choices have a direct impact on productivity and quality of output.

The incorporation of AI into industrial cobots has several practical applications. Businesses may get an advantage over their rivals thanks to AI-enabled cobots, which are more accurate, safer, scalable, and intelligent. Therefore, it's expected that artificial intelligence will be widely used in cobots in the near future.

## 5.8 IMPACT OF AI ON THE DESIGN AND FUNCTION OF COBOTS

AI is having an ever-increasing effect on the design and functionality of cobots as organizations across the globe increasingly embrace them in their operations. Cobots are already being employed in industries as diverse as manufacturing, healthcare, agriculture, and retail because of advancements in AI. Until recently, cobots were strictly designed to carry out single tasks, with no leeway for responding to unexpected events. As AI has progressed, cobots have gained the ability to learn from their experiences and make judgments based on the information available to them. This is especially helpful in dynamic settings, where conditions might rapidly change and it would be impracticable to manually retrain the robot to deal with the new circumstances. Cobot security is another area where AI is being used. Cobots can now identify and avoid possible dangers because of AI's capacity to teach robots to recognize their surroundings. As a result, cobots and humans may collaborate with fewer hurdles and more ease. Cobots are also improving in their ability to do difficult jobs thanks to AI. Cobots may be taught to do activities more correctly and effectively with the help of AI-driven machine learning. This is especially helpful in healthcare, where a cobot's precision and swiftness may have a significant impact on the quality of treatment provided to patients. Cobots' form and function will be heavily influenced by AI as they get toward human levels of sophistication. By enabling cobots to respond to changing conditions and complete complex tasks, AI is making cobots more capable and versatile than ever before. This is revolutionizing the way businesses operate and has the potential to drastically improve productivity and efficiency.

## 5.9 STANDARDS AND SAFETY FOR COBOTS AND ROBOTS

The international standard ISO 10218 and the Technical Specification ISO/TS 15066: 2016, the American ANSI/RIA R15.06, the European EN 775 that is adapted from ISO 10218, and standards like the Spanish UNE-EN 755 that is adapted from EN 775 by the Spanish Association of Standardization and Certification all account for the risks associated with the use of collaborative robots by workers. The examination of these hazards should inform the choice of a safety system to reduce the likelihood of accidents. In the past, it was common practice for security systems to install barriers between human and robotic workers. The UNE-EN 755: 1996 standard is one example of a document that reflects this split. It recommended installing sensors to alert personnel to danger in an area where the robotic system's current status might pose a threat to their safety. Conventional wisdom holds that only when a robot is not operating autonomously may authorized persons enter its workplace. The most recent versions of ISO 10218-1 and ISO 10218-2 outline the necessary conditions for collaborative work and classify different kinds of cooperative endeavors. As an example of the former, we can think of manual guiding, an interface window, and a collaborative workspace; as an example of the latter, we can think of start-up controls, operation of the safety control system, motion braking, and speed control. The terminology for robots and robotic equipment is defined in ISO: 8373-2012, an international standard. New words, such as human–robot interaction and the service robot, and more established terms, such as robot and control system, are developed to help with the creation of new collaborative activities in industrial and nonindustrial situations. ISO/ TS 15066:2016 is an updated technical definition that aims to establish human–robot cooperation by expanding on the criteria and recommendations made in ISO 10218.

## 5.10   POTENTIAL FOR AI TO ENHANCE THE SAFETY OF COBOTS

The robotics industry is fast becoming interested in the potential of AI to increase the security of collaborative robots. There is a pressing need to guarantee the safe and effective operation of robots due to their widespread use in manufacturing and other sectors. By giving these robots more command over their surroundings and allowing for more precise risk assessment, AI has the potential to radically improve the security of collaborative robots. Artificial intelligence may be used to spot dangers in the environment and warn humans so they can intervene. AI can be used to keep the robot's operator apprised of any potential threats in the area, monitor the robot's actions for any signs of abnormality, and notify the user as needed. The cobot's operating environment may also be managed more precisely with the help of AI. The presence of hazardous compounds, adjacent moving objects, and other situations are only some of the aspects that may be taken into consideration by AI to determine the level of danger present in the surrounding area. Because of this, the robot will be able to function in a risk-free and productive environment. AI may also be used to accurately foresee the dangers that the robot's actions will expose it to. The robot can keep tabs on its environment and respond appropriately. AI can almost certainly make collaborative robots safer to use. AI has the potential to give robots more control over their environments, enable them to make more accurate risk assessments, and allow humans to keep a closer eye on their whereabouts and activities. In terms of both security and productivity, this might revolutionize the employment of collaborative robots.

## 5.11   ROLE OF AI IN COMMUNICATION BETWEEN HUMANS AND COBOTS

AI is an essential part of the human–cobot dynamic. It improves the robot's ability to comprehend and act on human requests. Cobots can recognize things and carry out activities like sorting with the assistance of artificial intelligence. Artificial intelligence also contributes to job safety. Cobots may be set up to monitor their surroundings and cease operating if they see a human. As a result, fewer injuries and accidents will occur. Artificial intelligence may also be used to spot programming mistakes in a cobot, warning humans of potential dangers ahead of time. Finally, AI allows cobots to pick up new skills from their human handlers. The efficiency and accuracy of cobots may be improved by training them to recognize patterns and predict human movements. This increases cobots' productivity and efficiency in the workplace, which in turn may save costs and boost the quality of service provided to customers. AI plays a pivotal role in the seamless collaboration between humans and cobots. AI aids the efficiency and effectiveness of cobots in the workplace by allowing them to comprehend and react to human directions, identify possible faults, and learn from experience.

## 5.12  SCOPE AND CONCLUSIONS

Cobots, which facilitate human–robot cooperation, are another promising technology for accommodating the rising complexity and adaptability of the industrial sector. Some of the most important facilitators of cobots' intelligent and adaptable collaboration with human operators are tools for programming that are both intuitive and human aware. Recent years have seen a flurry of studies in this area. In this work, we compile the most recent findings on cobot programming. First, the programming needs and situations for cobot implementations in the industrial setting are outlined. Then, the cobot programming technologies are broken down into three categories: communication, optimization, and learning. Finally, the literature sourrounding these three subcategories is thoroughly examined. To facilitate cooperation, cobots are equipped with communication tools that allow a human operator to convey their intentions or directions to the robot. Online developers create algorithms called optimization features, which allow a cobot to learn from interactions with a collaborative operator and adjust its behavior accordingly, all following an optimal policy model. A cobot with learning capabilities can figure out its policy with a little help from its human collaborator. The article discusses how aspects like communication, optimization, and learning take into account human intuition and consciousness. Additionally, research needs for cobot programming are compared to the current state of the art. Finally, suggestions for further study and implementation of cobot programming are made to further facilitate collaborative situations in the industrial setting.

Collaborative robots that use artificial intelligence are becoming more vital to today's manufacturing sector. The increasing intelligence of robots means that they will soon be able to do a wide variety of human jobs, from manufacturing to healthcare. Robots powered by artificial intelligence have enormous promise, but there are still several obstacles in the way of their widespread adoption in the workplace. Robots powered by artificial intelligence must, first and foremost, properly perceive and react to their surroundings. For this purpose, sophisticated algorithms that can process and synthesize information from a variety of sources at high speeds are needed. In addition, these algorithms need to be flexible enough to respond to new circumstances. Since human behavior is notoriously unpredictable and prone to sudden shifts, this is an extremely challenging challenge when dealing with human–robot interactions. Second, robots powered by AI need to be able to work in tandem with people. To achieve this goal, it is necessary to create machines that can comprehend and respond to the requirements of their human counterparts. In addition, robots need the capability to reflect on their actions and make corrections. These robots can interact with people in a way that prevents harm to either party. Before AI-driven robots can be completely incorporated into the workplace, these problems provide a daunting set of barriers to overcome. These obstacles must be surmounted before the revolutionary promise of AI-driven robots in the contemporary workplace can be realized, but this is entirely possible with the appropriate mix of research and development.

## REFERENCES

[1] Bi ZM, Lang SYT. A framework for CAD- and sensor-based robotic coating automation. *IEEE Trans Ind Informatics* [Internet]. 2007;3:84–91. Available from: https://ieeexplore.ieee.org/document/4088938/.

[2] Ratasich D, Khalid F, Geissler F, et al. A roadmap toward the resilient Internet of Things for cyber-physical systems. *IEEE Access* [Internet]. 2019;7:13260–13283. Available from: https://ieeexplore.ieee.org/document/8606923/.

[3] Garcia MAR, Rojas R, Gualtieri L, et al. A human-in-the-loop cyber-physical system for collaborative assembly in smart manufacturing. *Procedia CIRP* [Internet]. 2019;81:600–605. Available from: https://linkinghub.elsevier.com/retrieve/pii/S2212827119304676.

[4] Valori M, Scibilia A, Fassi I, et al. Validating safety in human-robot collaboration: Standards and new perspectives. *Robotics* [Internet]. 2021;10:65. Available from: https://www.mdpi.com/2218-6581/10/2/65.

[5] Wong C, Yang E, Yan X-T, et al. Autonomous robots for harsh environments: A holistic overview of current solutions and ongoing challenges. *Syst Sci Control Eng* [Internet]. 2018;6:213–219. Available from: https://www.tandfonline.com/doi/full/10.1080/21642583.2018.1477634.

[6] Dzedzickis A, Subačiūtė-Žemaitienė J, Šutinys E, et al. Advanced Applications of industrial robotics: New trends and possibilities. *Appl Sci* [Internet]. 2021;12:135. Available from: https://www.mdpi.com/2076-3417/12/1/135.

[7] Vaidya S, Ambad P, Bhosle S. Industry 4.0 - A glimpse. *Procedia Manuf* [Internet]. 2018;20:233–238. Available from: https://linkinghub.elsevier.com/retrieve/pii/S2351978918300672.

[8] Galin R, Meshcheryakov R. Automation and robotics in the context of Industry 4.0: The shift to collaborative robots. *IOP Conf Ser Mater Sci Eng* [Internet]. 2019;537:032073. Available from: https://iopscience.iop.org/article/10.1088/1757-899X/537/3/032073.

[9] Nair VV, Kuhn D, Hummel V. Development of an easy teaching and simulation solution for an autonomous mobile robot system. *Procedia Manuf* [Internet]. 2019;31:270–276. Available from: https://linkinghub.elsevier.com/retrieve/pii/S235197891930407X.

[10] Wang YQ, Hu YD, El Zaatari S, et al. Optimised learning from demonstrations for collaborative robots. *Robot Comput Integr Manuf* [Internet]. 2021;71:102169. Available from: https://linkinghub.elsevier.com/retrieve/pii/S0736584521000533.

[11] Vysocky A, Novak P. Human-robot collaboration in industry. *MM Sci J* [Internet]. 2016;2016:903–906. Available from: https://www.mmscience.eu/june-2016.html#201611.

[12] Mittal S, Khan MA, Romero D, et al. A critical review of smart manufacturing & industry 4.0 maturity models: Implications for small and medium-sized enterprises (SMEs). *J Manuf Syst* [Internet]. 2018;49:194–214. Available from: https://linkinghub.elsevier.com/retrieve/pii/S0278612518301341.

[13] BMW Group. Innovative human-robot cooperation in BMW group production [Internet]. 2013. Available from: https://www.press.bmwgroup.com/global/article/detail/T0209722EN/innovative-humanrobot-%0Acooperation-in-bmw-group-production?language=en.

[14] El Makrini I, Elprama SA, Van den Bergh J, et al. Working with Walt: How a cobot was developed and inserted on an auto assembly line. *IEEE Robot Autom Mag* [Internet]. 2018;25:51–58. Available from: https://ieeexplore.ieee.org/document/8360084/.

[15] Schou C, Andersen RS, Chrysostomou D, et al. Skill-based instruction of collaborative robots in industrial settings. *Robot Comput Integr Manuf* [Internet]. 2018;53:72–80. Available from: https://linkinghub.elsevier.com/retrieve/pii/S0736584516301910.

[16] KUKA. Many wrenches make light work: KUKA flexFELLOW will provide assistance during drive train pre-assembly [Internet]. 2016 [cited 2022 Oct 20]. Available from: https://www.kuka.com/engb/%0Apress/news/2016/10/20160926vwsetztaufmensch-roboter-kollaboration.

[17] Robotics and Automation News. Innovative Skoda factory introduces human-robot collaboration with KUKA LBR IIWA [Internet]. 2017. Available from: https://roboticsandautomationnews.com/2017/02/16/innovative-skoda-factory-introduces-humanrobot-%0Acollaboration-with-kuka-lbr-iiwa/11404/.

[18] Alebooyeh M, Urbanic RJ. Neural network model for identifying workspace, forward and inverse kinematics of the 7-DOF YuMi 14000 ABB collaborative robot. *IFAC-PapersOnLine* [Internet]. 2019;52:176–181. Available from: https://linkinghub.elsevier.com/retrieve/pii/S2405896319308742.

[19] Winkelmann N. Human-robot cooperation at Audi [Internet]. 2017 [cited 2022 Oct 20]. Available from: https://www.springerprofessional.de/en/manufacturing/production---production-technology/humanrobot-%0Acooperation-at-audi/14221870.

[20] Universal Robots. UR10 cobots offer aging workforce solution and reduce relief worker costs for global car manufacturer [Internet]. 2018 [cited 2022 Oct 20]. Available from: https://www.universal-robots.com/casestories/%0Anissan-motor-company.

[21] El Zaatari S, Li W, Usman Z. Ring Gaussian mixture modelling and regression for collaborative robots. *Rob Auton Syst* [Internet]. 2021;145:103864. Available from: https://linkinghub.elsevier.com/retrieve/pii/S0921889021001494.

[22] El Zaatari S, Wang Y, Hu Y, et al. An improved approach of task-parameterized learning from demonstrations for cobots in dynamic manufacturing. *J Intell Manuf* [Internet]. 2022;33:1503–1519. Available from: https://link.springer.com/10.1007/s10845-021-01743-w.

[23] McGirr L, Jin Y, Price M, et al. Human robot collaboration: Taxonomy of interaction levels in manufacturing. *ISR Eur 2022, 54th Int Symp Robot* [Internet]. 2022. pp. 1–8. Available from: https://ieeexplore.ieee.org/abstract/document/9861817/authors#authors.

[24] Vysocky A, Novak P. Human - Robot collaboration in industry. *MM Sci J* [Internet]. 2016;2016:903–906. Available from: https://www.mmscience.eu/june-2016.html#201611.

[25] Deng S, Cai Z, Fang D, et al. Application of robot offline programming in thermal spraying. *Surf Coat Technol* [Internet]. 2012;206:3875–3882. Available from: https://linkinghub.elsevier.com/retrieve/pii/S0257897212002071.

[26] Ilavsky J, Allen AJ, Long GG, et al. Influence of spray angle on the pore and crack microstructure of plasma-sprayed deposits. *J Am Ceram Soc* [Internet]. 1997;80:733–742. Available from: https://onlinelibrary.wiley.com/doi/10.1111/j.1151-2916.1997.tb02890.x.

[27] Leigh SH, Berndt CC. Evaluation of off-angle thermal spray. *Surf Coat Technol* [Internet]. 1997;89:213–224. Available from: https://linkinghub.elsevier.com/retrieve/pii/S0257897296028976.

[28] Friis M, Persson C, Wigren J. Influence of particle in-flight characteristics on the microstructure of atmospheric plasma sprayed yttria stabilized ZrO2. *Surf Coat Technol* [Internet]. 2001;141:115–127. Available from: https://linkinghub.elsevier.com/retrieve/pii/S0257897201012397.

[29] Kang CW, Ng HW, Yu SCM. Imaging diagnostics study on obliquely impacting plasma-sprayed particles near to the substrate. *J Thermal Spray Technol* [Internet]. 2006;15:118–130. Available from: https://link.springer.com/10.1361/105996306X92686.

[30] Binder K, Gottschalk J, Kollenda M, et al. Influence of impact angle and gas temperature on mechanical properties of titanium cold spray deposits. *J Thermal Spray Technol* [Internet]. 2011;20:234–242. Available from: https://link.springer.com/10.1007/s11666-010-9557-1.

[31] Tillmann W, Vogli E, Krebs B. Influence of the spray angle on the characteristics of atmospheric plasma sprayed hard material based coatings. *J Thermal Spray Technol* [Internet]. 2008;17:948–955. Available from: https://link.springer.com/10.1007/s11666-008-9261-6.

[32] Mostaghimi J, Chandra S. Droplet impact and solidification in plasma spraying. In: Kulacki FA. editor, *Handbook of Thermal Science and Engineering.* Cham: Springer International Publishing; 2018. pp. 2967–3008. Available from: https://link.springer.com/10.1007/978-3-319-26695-4_78.

[33] Tillmann W, Baumann I, Hollingsworth P, et al. Influence of the spray angle on the properties of HVOF sprayed WC-Co coatings using (−10 + 2 μm) fine powders. *J Thermal Spray Technol* [Internet]. 2013;22:272–279. Available from: https://link.springer.com/10.1007/s11666-013-9882-2.

[34] Pukasiewicz AGM, de Boer HE, Sucharski GB, et al. The influence of HVOF spraying parameters on the microstructure, residual stress and cavitation resistance of FeMnCrSi coatings. *Surf Coat Technol* [Internet]. 2017;327:158–166. Available from: https://linkinghub.elsevier.com/retrieve/pii/S0257897217307752.

[35] Zhao L, Maurer M, Fischer F, et al. Influence of spray parameters on the particle in-flight properties and the properties of HVOF coating of WC-CoCr. *Wear* [Internet]. 2004;257:41–46. Available from: https://linkinghub.elsevier.com/retrieve/pii/S0043164803005775.

[36] Li M, Christofides PD. Computational study of particle in-flight behavior in the HVOF thermal spray process. *Chem Eng Sci* [Internet]. 2006;61:6540–6552. Available from: https://linkinghub.elsevier.com/retrieve/pii/S0009250906003769.

[37] Fauchais PL, Heberlein JVR, Boulos MI. *Thermal Spray Fundamentals* [Internet]. Boston, MA: Springer US; 2014. Available from: https://link.springer.com/10.1007/978-0-387-68991-3.

[38] Oberkampf WL, Talpallikar M. Analysis of a high-velocity oxygen-fuel (HVOF) thermal spray torch part 2: Computational results. *J Thermal Spray Technol* [Internet]. 1996;5:62–68. Available from: https://link.springer.com/10.1007/BF02647519.

[39] Stokes J. *Production of Coated and Free-Standing Engineering Components Using the HVOF (High Velocity Oxy-Fuel) Process.* Dublin, Ireland: Dublin City University; 2003.

[40] Stokes J, Looney L. Residual stress in HVOF thermally sprayed thick deposits. *Surf Coat Technol* [Internet]. 2004;177–178:18–23. Available from: https://linkinghub.elsevier.com/retrieve/pii/S0257897203009861.

[41] Sahraoui T, Guessasma S, Ali Jeridane M, et al. HVOF sprayed WC-Co coatings: Microstructure, mechanical properties and friction moment prediction. *Mater Des* [Internet]. 2010;31:1431–1437. Available from: https://linkinghub.elsevier.com/retrieve/pii/S0261306909004610.

[42] Yilbas BS, Al-Zaharnah I, Sahin A. *Flexural Testing of Weld Site and HVOF Coating Characteristics* [Internet]. Berlin, Heidelberg: Springer Berlin Heidelberg; 2014. Available from: https://link.springer.com/10.1007/978-3-642-54977-9.

[43] Thirumalaikumarasamy D, Shanmugam K, Balasubramanian V. Influences of atmospheric plasma spraying parameters on the porosity level of alumina coating on AZ31B magnesium alloy using response surface methodology. *Prog Nat Sci Mater Int* [Internet]. 2012;22:468–479. Available from: https://linkinghub.elsevier.com/retrieve/pii/S1002007112000949.

[44] Kar S, Paul S, Bandyopadhyay PP. Processing and characterisation of plasma sprayed oxides: Microstructure, phases and residual stress. *Surf Coat Technol* [Internet]. 2016;304:364–374. Available from: https://linkinghub.elsevier.com/retrieve/pii/S0257897216306351.

[45] Wang Y, Li KY, Scenini F, et al. The effect of residual stress on the electrochemical corrosion behavior of Fe-based amorphous coatings in chloride-containing solutions. *Surf Coat Technol* [Internet]. 2016;302:27–38. Available from: https://linkinghub.elsevier.com/retrieve/pii/S0257897216304182.

[46] Nazir MH, Khan ZA, Stokes K. Analysing the coupled effects of compressive and diffusion induced stresses on the nucleation and propagation of circular coating blisters in the presence of micro-cracks. *Eng Fail Anal* [Internet]. 2016;70:1–15. Available from: https://linkinghub.elsevier.com/retrieve/pii/S1350630716300978.

[47] Tian H, Wang C, Guo M, et al. A residual stresses numerical simulation and the relevant thermal-mechanical mapping relationship of Fe-based coatings. *Results Phys* [Internet]. 2019;13:102195. Available from: https://linkinghub.elsevier.com/retrieve/pii/S2211379719300129.

[48] Gadow R, Killinger A, Martinez V. 4 Glass and glass ceramic layer composites with functional coatings. In: Gadow R, Mitic VV, editor. *Advanced Ceramics and Applications* [Internet]. Berlin: De Gruyter. 2021. pp. 41–60. Available from: https://www.degruyter.com/document/doi/10.1515/9783110627992-004/html.

[49] Gustavsson P, Syberfeldt A, Brewster R, et al. Human-robot collaboration demonstrator combining speech recognition and haptic control. *Procedia CIRP*. 2017;63:396–401.

[50] Gabler V, Stahl T, Huber G, et al. A game-theoretic approach for adaptive action selection in close proximity human-robot-collaboration. *2017 IEEE Int Conf Robot Autom.* IEEE; 2017. pp. 2897–2903.

[51] Munzer T, Toussaint M, Lopes M. Efficient behavior learning in human-robot collaboration. *Auton Robots.* 2018;42:1103–1115.

[52] Vogt D, Stepputtis S, Grehl S, et al. A system for learning continuous human-robot interactions from human-human demonstrations. *2017 IEEE Int Conf Robot Autom.* IEEE; 2017. pp. 2882–2889.

[53] Frutos A. Numerical Analysis of the Temperature Distribution and Offline Programming of Industrial Robot for Thermal Spraying [Internet]. Stuttgart, Germany: IMTCCC University of Stuttgart; 2009. Available from: https://repositorio.upct.es/bitstream/handle/10317/764/pfc2966.pdf;jsessionid=0F747A3FBF744AC00A88B41068B0142D?sequence=1.

[54] Kout A, Müller H. Parameter optimization for spray coating. *Adv Eng Softw* [Internet]. 2009;40:1078–1086. Available from: https://linkinghub.elsevier.com/retrieve/pii/S0965997809000611.

[55] Trifa F-I, Montavon G, Coddet C. On the relationships between the geometric processing parameters of APS and the $Al_2O_3$-$TiO_2$ deposit shapes. *Surf Coat Technol* [Internet]. 2005;195:54–69. Available from: https://linkinghub.elsevier.com/retrieve/pii/S0257897204007108.

[56] Guessasma S, Bounazef M, Nardin P. Neural computation analysis of alumina-titania wear resistance coating. *Int J Refract Met Hard Mater* [Internet]. 2006;24:240–246. Available from: https://linkinghub.elsevier.com/retrieve/pii/S0263436805000788.

[57] Candel A, Gadow R. Trajectory generation and coupled numerical simulation for thermal spraying applications on complex geometries. *J Thermal Spray Technol* [Internet]. 2009;18:981–987. Available from: https://link.springer.com/10.1007/s11666-009-9338-x.

[58] Fang D, Deng S, Liao H, et al. The effect of robot kinematics on the coating thickness uniformity. *J Thermal Spray Technol* [Internet]. 2010;19:796–804. Available from: https://link.springer.com/10.1007/s11666-010-9470-7.

[59] Hansbo A, Nylén P. Models for the simulation of spray deposition and robot motion optimization in thermal spraying of rotating objects. *Surf Coat Technol* [Internet]. 1999;122:191–201. Available from: https://linkinghub.elsevier.com/retrieve/pii/S0257897299002558.

[60] Zha XF. Optimal pose trajectory planning for robot manipulators. *Mech Mach Theory* [Internet]. 2002;37:1063–1086. Available from: https://linkinghub.elsevier.com/retrieve/pii/S0094114X02000538.

[61] Zhang XC, Xu BS, Wang HD, et al. Modeling of the residual stresses in plasma-spraying functionally graded ZrO2/NiCoCrAlY coatings using finite element method. *Mater Des* [Internet]. 2006;27:308–315. Available from: https://linkinghub.elsevier.com/retrieve/pii/S0261306904002845.

[62] Ng HW, Gan Z. A finite element analysis technique for predicting as-sprayed residual stresses generated by the plasma spray coating process. *Finite Elem Anal Des* [Internet]. 2005;41:1235–1254. Available from: https://linkinghub.elsevier.com/retrieve/pii/S0168874X05000387.

[63] Amara M, Timchenko V, El Ganaoui M, et al. A 3D computational model of heat transfer coupled to phase change in multilayer materials with random thermal contact resistance. *Int J Thermal Sci* [Internet]. 2009;48:421–427. Available from: https://linkinghub.elsevier.com/retrieve/pii/S1290072908000598.

[64] Khor K., Gu Y. Effects of residual stress on the performance of plasma sprayed functionally graded ZrO2/NiCoCrAlY coatings. *Mater Sci Eng A* [Internet]. 2000;277:64–76. Available from: https://linkinghub.elsevier.com/retrieve/pii/S0921509399005651.

[65] Bolot R, Planche M-P, Liao H, et al. A three-dimensional model of the wire-arc spray process and its experimental validation. *J Mater Process Technol* [Internet]. 2008;200:94–105. Available from: https://linkinghub.elsevier.com/retrieve/pii/S0924013607007807.

[66] El Zaatari S, Wang Y, Li W. Reinforcement learning based optimization for cobot's path generation in collaborative tasks. *2021 IEEE 24th Int Conf Comput Support Coop Work Des* [Internet]. IEEE; 2021. pp. 975–980. Available from: https://ieeexplore.ieee.org/document/9437874/.

[67] Bogue R. The growing use of robots by the aerospace industry. *Ind Robot An Int J* [Internet]. 2018;45:705–709. Available from: https://www.emerald.com/insight/content/doi/10.1108/IR-08-2018-0160/full/html.

[68] Cutting Tool Engineering. Maintenance-free cobots operate continuously in harsh environment. *CTE MAG* [Internet]. 2017. Available from: https://www.ctemag.com/news/industry-news/maintenance-free-cobots-operate-continuously-harsh-environment.

# 6 Challenges of Using Artificial Intelligence in Thermal Spray Industry

## Implementation, Optimization, and Control

### ABBREVIATIONS

| | |
|---|---|
| **AI** | Artificial intelligence |
| **CMM** | Coordinate measuring machine |
| **Cobots** | Collaborative robots |
| **CV** | Cross-validation |
| **HEA** | High-entropy alloys |
| **HVOF** | High-velocity oxy-fuel |
| **LDA** | Linear discriminant analysis |
| **ML** | Machine learning |
| **MSE** | Mean square error |
| **NMF** | Non-negative factorization |
| **PCA** | Principal component analysis |
| **RF** | Random forest |
| **RMSE** | Mean absolute error root mean square error |
| **R2** | Coefficient of determination |
| **SME** | Small and medium-sized business |
| **SOD** | Stand-off distance |
| **TS** | Thermal spraying |

### 6.1 BARRIERS TO ARTIFICIAL INTELLIGENCE (AI) IMPLEMENTATION

The coatings industry is undergoing digitalization, much like the rest of the manufacturing industries. Every aspect of a company, from brainstorming to manufacturing to customer service, stands to benefit from the exponential growth in the availability of useful information. The challenge is finding out where and how to use techniques like artificial intelligence (AI) and natural language processing to make the available data related to the problem or topic of interest. Although the future of AI has not yet been well-defined, there are many challenges to overcome. Machine learning (ML) techniques are renowned for their capacity to learn features from data representations

 DOI: 10.1201/9781003400660-6

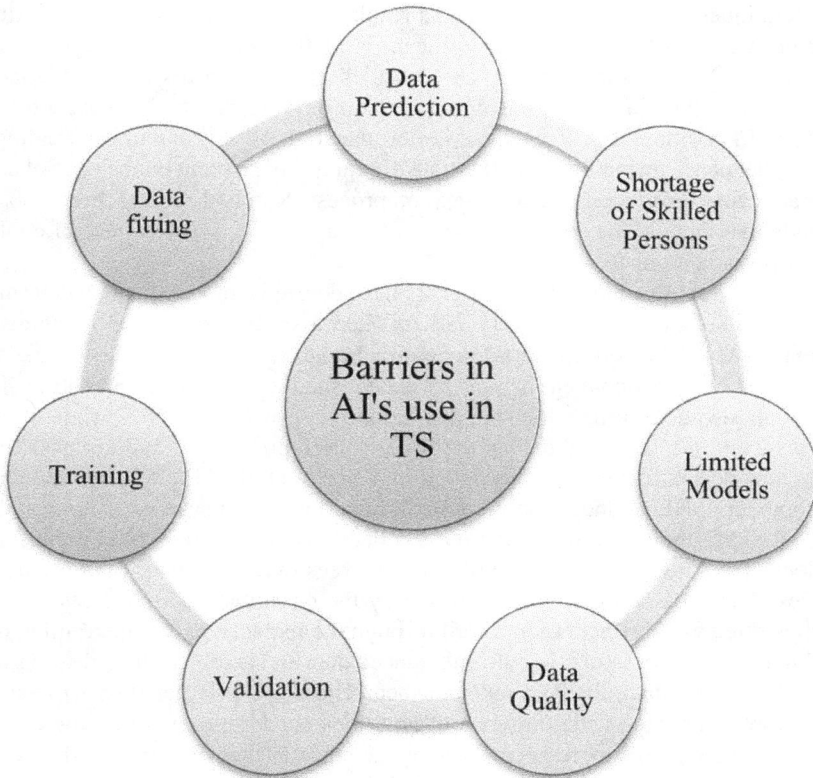

**FIGURE 6.1**   Barriers or challenges in artificial intelligence's use in thermal spray.

using complicated neural network designs to approximate nonlinearity. But certain barriers come in the way of adopting AI by the coating industry. In Figure 6.1, a schematic illustration of barriers and challenges to AI's use in the thermal spray (TS) industry is given. In the following sections of this chapter, various barriers to the implementation of AI in TS are addressed.

## 6.2   DATA PREDICTION

The use of high-entropy alloys (HEA) in creating surfaces specifically suited to the unique requirements of a given application is a game-changing opportunity in the realm of surface engineering and modern coatings. HEAs give producers new possibilities for producing substitutes for expensive, rare, and toxic materials that are subject to international regulations. They are also built to work around the constraints of existing materials that are part of a thirty-year-old legacy. Several studies have demonstrated encouraging results for calculating phases of such complicated systems with the latest HEA data build-up [1–5]. But pure regularization and overfitting in the field of high-entropy TS alloy coatings are two current drawbacks faced by

ML techniques. Another issue is that the results of neural networks are sometimes difficult to comprehend. Another significant issue that reduces the efficacy of ML algorithms when constructing the TS feedstock HEA materials is the absence of information in the utmost datasets detailing their development and the needed cooling rates. In addition, it is of the most importance to have a strong understanding of the fact that phase formation from the state of liquid is dependent on the pace of cooling that occurs throughout the solidification process. Rarely do researchers resort to methods that include the phase change of alloys and depend on the pace of cooling in their investigations [6].

To circumvent this obstacle, researchers have devised a model that can determine which solid solution phase will crystallize when a specific elemental combination is cooled during the solidification process of liquid in water atomizers or gas and then chilled once again when the powder is deposited to generate a TS coating. This model can also determine which solid solution phase will crystallize when a given elemental mixture is cooled during the solidification process of liquid in gas. To do this, a new design feature was developed by Kamnis et al. [7] to forecast the phase development of HEAs under continuous solidification circumstances using a random forest (RF) ML model. In an RF, there is a collection of decision trees. For non-linear relationships or noisy data sets, single decision trees frequently have poor prediction abilities. These flaws are addressed in RF by the ensemble decision trees, each of which is fitted to a distinct random trail through the replacement of the training data set. Since it can successfully handle unbalanced data and is robust when dealing with high-dimensional data, the RF model has been chosen. The larger class will experience a low error rate in the presence of an unbalanced data set, while the smaller class will experience a higher error percentage since RF seeks to reduce the overall error rate. Previous research has shown its improved performance in various materials science disciplines, where data are frequently scarce [8]. The model properly predicted the phase development in almost all of the coated alloys and the material that was atomized to make them.

## 6.3  SHORTAGE OF SKILLED PERSONS

Misuse of these cutting-edge instruments by inexperienced users is a different problem. Many individuals still lack the necessary expertise to make effective use of these technologies. An incorrect application usually yields incorrect outcomes, which calls into question the usefulness of such programs [9]. TS Companies in every industry are realizing that it is difficult to locate and employ talented data scientists and AI professionals. For AI projects, an interdisciplinary team consisting of data scientists, ML developers, and software architects is required to collaborate. Many businesses either do not have access to these resources or do not have the financial means to make use of them for even a single data science project. It must be remembered that well-trained personnel's practical expertise is a crucial resource for process control and diagnostics both today and in the future. Therefore, efficient means of knowledge collection, systematization, representation, and transfer will be crucial.

## 6.4  DATA QUALITY AND QUANTITY

It is essential to the success of AI projects to have access to data that is uncontaminated, relevant, and of high quality; yet, this might be difficult to achieve on the shop floor during coating deposition. The collection of data almost always involves some degree of error, the origin of which might lie in any one of several different factors. One example is sensor data that was obtained under challenging circumstances, such as when there was an excessive amount of noise or vibration that might have contributed to inaccurate readings. Another challenge in AI is that AI predictions are much better suited for the big dataset than the smaller dataset [9]. The amount and quality of available data are likely to pose significant difficulties. One hundred to two hundred data points for experimental formulations may exist in many firms' data sets. That's hardly big data, and it certainly isn't enough to make good use of ML and AI, which generally transmit hundreds or millions of data points. To make the most of limited information, it is necessary to create intermediate solutions. Another distinct problem arises when untrained individuals misuse such sophisticated equipment. The widespread lack of proficiency in the use of such technology is a problem in the present day. An incorrect application usually yields incorrect outcomes, which in turn casts doubt on the success of the project.

## 6.5  LIMITED MODELS AND THEIR VALIDATION FOR PROCESS DIAGNOSTICS

Models are crucial tools for understanding how complex the various TS techniques are in terms of fluid flow properties, phase changes, heat, momentum transfer, etc. To do so would need more precise and improved stochastic models. Future spray process control improvements will be made possible in part by feeding numerical models into ML techniques. To automate the process of creating robot trajectories, Cai et al. [10] suggest developing a new software package for the offline programming tool RobotStudioTM that takes into consideration a very simple coating model that considers the kinematic characteristics of the torch. A mathematical model for the heat transfer experienced by plasma-coated cylinders is proposed by Ding et al. [11]. Numerous numerical methods are given for simulating splat creation in TS [12–19]. Although these methods are essential for understanding the physics at play, they are not well suited for full-scale simulations of coating deposition. The ML-driven Self-Organizing-Map technique seems to be highly successful in application optimization via automated robot planning and generates fantastic outcomes when compared to those produced by a human expert. The primary challenge with this method is how efficiently it performs when applied to non-axisymmetric 3D objects. Hence, more study is necessary to create a full multidimensional model.

The methods for particle diagnostics in standard TS methods are unsuitable since they can only be utilized for long spray distances and detect only a particle portion, if any at all [20,21]. In a study, Akbarnozari et al. [22] used laser scattering and Mie's scattering theories to study the Gaussian-distributed approach to study suspension particle size in the plasma jet. The smallest size of the suspension particles is around 5 µm, as this is the size at which traditional particle analysis techniques reach their imaging and resolution limits. Key parameters for plasma generation, suspensions, solutions, and injectors could rarely be specifically optimized due to the limited application of these diagnostic procedures and have instead mostly been developed on the basis of empirical research.

In novel spray processes like cold spray, warm spraying, and high-velocity air fuel spraying, low particle temperatures provide further special difficulty for process diagnostics. In this case, the thermal radiation power is too weak to be detected and valued using Planck's law based on two-color pyrometry. Consequently, it is difficult to measure the particle's thermal condition both during flight and at impact. To determine and evaluate the proper plastic deformation model for collisions of particles at low temperatures or close to their melting points, more research is required. Once more, since these striking particles are the foundation of the coatings, their attributes directly affect those of the coatings.

Forecasting the characteristics and microstructure of a TS coating based on understanding the in-flight properties of the flow of particles, torch movements, form, and the substrate's roughness and temperature throughout spraying is still a difficult task. It is difficult to get a complete understanding of the "simple" effect that a molten plasma spray droplet has on a smoother substrate when the temperature is at room temperature. When the substrate is heated to temperatures between 300°C and 400°C, droplet impact numerical models can provide accurate predictions about the uniform smoothing, and solidification processes of the impacting particle droplet, which is in better thermo-mechanical interaction with the underlying substrate. The droplet's spreading behavior drastically transforms when vapor of water or other gases is absorbed on the substrate at normal temperature. Liquid makes complete contact with the substrate just at the moment of impact, but as it expands, it pulls away and produces a thin, growing molten film that cracks when it's too thin. Droplet dispersion and solidification at room temperature have been surprisingly poorly represented by models. Since plasma spray coatings are composed of individual droplets that have flattened and hardened, often known as splats or lamellae, this is a crucial area of study.

The above-discussed models' validation provides a significant challenge that demands additional consideration. For instance, the difficulty of measuring particle temperature and size under suspension plasma spray circumstances makes it challenging to estimate in-flight particle characteristics. Furthermore, ML models are only as reliable as the information and assumptions upon which they are based. If the assumptions used to train a model are flawed, then the model will likely provide inaccurate results and exhibit unpredictable behavior, regardless of how much data was utilized in the training process.

## 6.6   TRAINING AND VALIDATION OF MODELS

There are a few challenges in the training and validation of the models, as illustrated in Figure 6.2. The disadvantage of a data-driven modeling method is the necessity of a large quantity of process training data to obtain a high degree of prediction accuracy, which was also highlighted in the research that was carried out. This is something that has been recently noted in relevant studies about manufacturing [23,24].

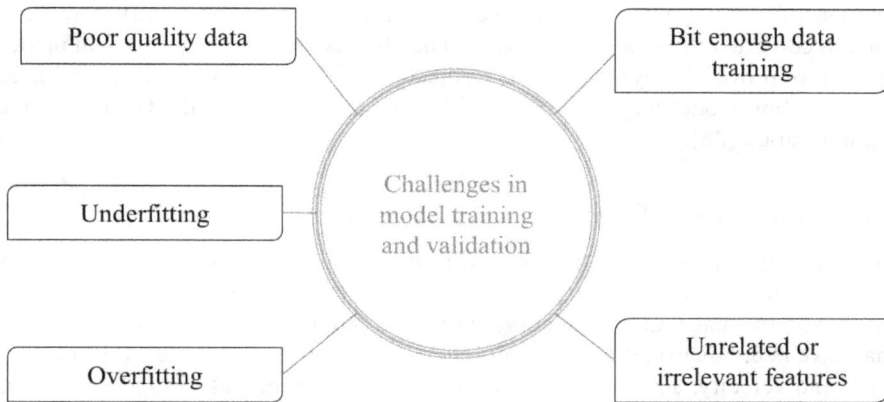

FIGURE 6.2   Challenges in model training and validation.

High experimental expenses and the absence of an automated measuring method in high production rate additive manufacturing are related to this data scarcity problem. To solve the problem, For the first time, Liu et al. [23] combined numerical and data-driven modeling techniques to use a gray modeling method in a TS process. Despite the fair forecast accuracy attained in this study, the scientists concluded that more complicated and non-linear events occurred. They recommended additional investigation of data-efficient modeling techniques to increase prediction accuracy.

## 6.7   POSSIBLE SOLUTIONS TO TACKLE CHALLENGES IN AI IMPLEMENTATION IN THE TS INDUSTRY

A possible solution to tackle challenges in AI implementation in the TS industry is the use of the basics during the processing of data, algorithms, validation, training, tuning, fitting, and testing. The following steps can be accurately carried out to resolve such challenges in the TS industry.

### 6.7.1   PROBLEM FORMULATION

To put it simply, AI applications cannot work without prediction models, where a model is a specified set of different rules used by the AI algorithm to help in learning autonomously [25]. The computer program improves its performance on certain tasks as a result of its learning, while it remains rather static on others. Many types of interactions and activities fall under its remit, and it may accommodate different measures of success. This essay, however, focuses on supervised learning via concrete examples and complicated tasks of varied degrees. Finally, we considered how both assignments and results might have an impact. Point characteristics, denoted by $x$, are located in the $n$-dimensional space $Rn$ (where $n$ is the number of dimensions) [26,27]. In linear regression, these factors are known as independent variables. These characteristics are categorized as objectives. Different AI algorithms are fed the stated features to discover the relationships between the features and the targets. Training the neural network is used to construct the prediction models. The effectiveness of the prediction model may be improved by hyperparameter adjustment. The development of a reliable AI prediction model may also be aided by the use of other methods, such as L2 regularization [28].

### 6.7.2  PROCESSING A DATASET

Preprocessing approaches, such as normalization and standardization, may help make the dataset more manageable before processing [25–27,29–32]. Variety, value, volume, pace, and authenticity are some of the characteristics of a dataset that have been identified by research [25,32]. The 3-V qualities, i.e., variety, volume, and velocity, apply to the data production, capture, and storage processes, and they characterize the "veracity" and "value" elements that are necessary to acquire useful and relevant information from the input. For this article, the phrase "big data" has some significance if and only if it meets an expanded version of each of the five specified definitions. Large, complex amounts of information that need "intelligent methods" for analysis are what "big data" refers to in the technical sense. The location of each data component and substance is precisely known in structured data, which is knowledge with a predetermined data model [25,32,33]. Semi-structured data, also known as unstructured data, is a kind of data that does not follow the organizational framework of tables or data but divides structural parts into recognizable parts and characteristics into hierarchies [25,33]. In other cases, the data lacks structure or does not conform to predetermined models [25,33].

In this context, information may be categorized as either structured, semi-structured, or unstructured. A dataset based on data points is structured data, and a dataset based on images or videos is known as unstructured data. Quantitative data from unstructured data in the form of photographs is used to verify continuity and quality, as will be described in the next section. To aid in quality control, trends and model parameters may be distinguished using the structured data gleaned through mechanical studies and simulations. An effective solution may be achieved by appropriately blending data and structure [25–27,29,30]. A choice must be made about the nature of the issue, the nature of the task, the diversity, truthfulness, and amount of data needed on a case, and the functional implementation of ML.

### 6.7.3  FINDING THE MOST IMPORTANT FACTORS AND SIMPLIFYING THE DATA SET

Although feature creation for structured data is relatively easy [31], feature engineering [34] is an essential step when working with unstructured data. Current, extremely effective methods may be used to deal with the dimensionality of data; however, having too many characteristics can lead to complexity, earning it the nickname "the curse of dimensionality" [25–27]. Dimensional reduction and feature selection methods may be used to address the issue. Additional details may be found in the aforementioned papers [31,35]. Feature engineering is an important technique in both supervised and unsupervised learning for determining which data points are most important (microstructure, temperature, etc.) [34]. The user is responsible for determining which of the system's many possible variables has the greatest impact on performance. Function engineering refers to this process, which may be expanded upon by analyzing the statistical connection between input and output [36]. A correlation matrix computation is one such assessment approach that may be used to find important inputs and quantify their attributes [36]. Even if numerous input variables were employed for each data point, the computational efficiency of ML models would suffer in such high-dimensional input spaces [37]. Model training may benefit from dimensionality reduction strategies like Non-negative factorization (NMF) [38], Linear discriminant analysis (LDA) [37], and Principal component analysis (PCA) [39]. In a nutshell, the idea behind these techniques is to replace or supplement the original inputs with information derived from a variety of current sources [34,37]. These methods may enhance the computational capacity or efficiency of a ML model by making use of the fact that it is feasible to incorporate many inputs into a single model's dimensional metrics. To train a ML model, input has to have more than five points. The dimensionality, or number of data points, in the training set grows proportionally with the size of the dataset. The nature of the ML algorithm and the desired outcome both contribute to its growth.

### 6.7.4 Model and Loss Function

When discussing the predicted data processing of an AI, we use the letter '$T$' as an assignment. Labeling research specimens so that their completeness or destruction may be determined is one function of assignment $T$. A useful predictor of how well an AI system will perform on the job at hand is success in the task evaluation measure $P$. Improvements in classification accuracy, as indicated above, might be seen as "a potential measure" [29–31]. It's not as simple as plugging numbers into a formula to get the best metrics for success. Given the data-centric nature of this research, the T-assignment can only be used for model-based problems. The AI applies a novel mathematical approach to the problem of improving performance on the measure $P$. The symbol $P$ computes a gain that may be optimized for, making it the objective, cost, or loss function in many ML and DL scenarios. The term "objective function" is used throughout this piece to refer to the function we're trying to minimize or maximize from a statistical standpoint [29,30]. This is expressed as a minimum cost function $M$ in the context of a mathematical model of an AI system that includes factors and learns from the training of a dataset, $K$:

$$\alpha = \arg\min M\left(\theta | K\right) \qquad (6.1)$$

where the minimum value of arguments is denoted by arg min. Model validation often makes use of the loss functions detailed in Table 6.1. There is a strong correlation between the loss function and model performance [25,29,30]. Several loss functions always offer several optimal values for different tasks, but choosing which loss function to utilize before commencing a simulation exercise may be challenging.

**TABLE 6.1**

**Loss Functions that are often used in the Model Validation Process**

| S. No. | Model Name | Equation | Equation No. | Ref. |
|--------|------------|----------|--------------|------|
| 1 | Log loss | $$L = -\frac{1}{N} \sum_{j=1}^{N} \log\left(p_j\right)$$ | 6.2 | [40] |
| 2 | Mean square error (MSE) | $$MSE = \frac{1}{N} \sum_{j=1}^{N} \left(O_j - p_j\right)^2$$ | 6.3 | [40] |
| 3 | Mean absolute error (MAE) | $$MAE = \frac{1}{N} \sum_{j=1}^{N} \left|O_j - p_j\right|$$ | 6.4 | [41] |
| 4 | Mean absolute error root mean square error (RMSE) | $$RMSE = \sqrt{\frac{1}{N} \sum_{j=1}^{N} \left(O_j - p_j\right)^2}$$ | 6.5 | [41] |
| 5 | Coefficient of v($R^2$) | $$R^2 = 1 - \frac{E(S)}{T(S)}$$ | 6.6 | [42] |
| 6 | Mean absolute percent error (MAPE) | $$MAE = \frac{1}{N} \sum_{j=1}^{N} \frac{\left|O_j - p_j\right|}{p_j}$$ | 6.7 | [40] |

*The no. of observations (N), the predicted values ($p_j$), the actual values ($O_j$), the sum of squared errors (E(S)), and the total sum of squared errors (T(S)) are all shown here.*

## 6.7.5 DATA SPLITTING

After the data has been cleaned and visualized, many AI models are compared to one another. Glass designer's ultimate objective is to acquire a fully functional ML model with an extensive domain-independent understanding of the gathered data [29,30]. The final model must be able to predict future outcomes while maintaining a high degree of similarity to known outcomes. If the model is overfitted to the training data, it will not generalize well to new data [25–27,29,30]. Bias and variance are two separate but linked issues [25–27]. Methods such as leave-one-out cross-validation [35], *k*-cross-validation [43], holdout [44], and stratified *k*-fold cross-validation [40] are used to evaluate models in various ways [28].

## 6.7.6 UNDERFITTING AND OVERFITTING

Reduced model complexity leads to underfitting. The link between the inputs and the model's performance is not well explained by the model. In contrast, "noise" in the dataset is remembered to prevent overfitting [45]. When it comes to models, it is generally accepted that the greater the non-zero inputs, the greater the complexity

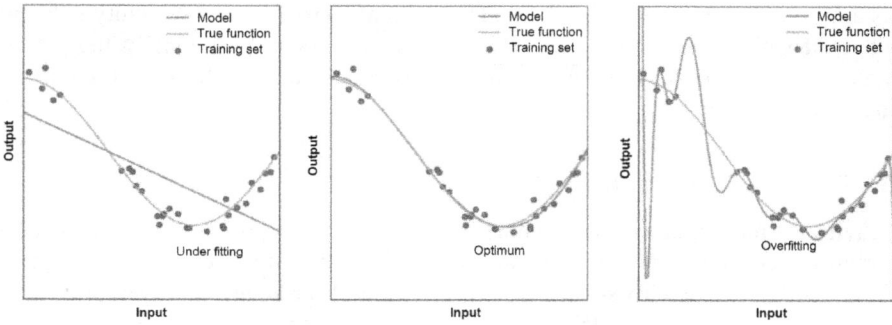

**FIGURE 6.3** Schematic representation of underfitting, optimum, and overfitting [46]. Permitted by authors to cite as permission.

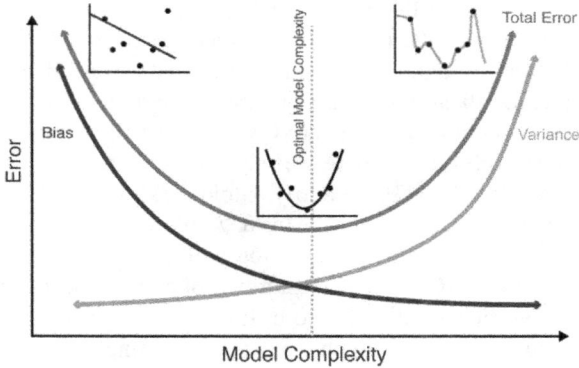

**FIGURE 6.4** Model complexity with respect to error [47]. Permissions from Elsevier.

of the built model [34]. Figure 6.3 depicts overfitting and underfitting as a function of training set size. Therefore, a $p = 1$ linear model cannot describe the relationship between input and output in this scenario. A model with $p = 15$ can adequately represent random data points within the training set, but it can't make many inferences about points outside the testing set (the validation set). Although eliminating noise from the dataset might not give adequate statistical power, the basic pattern of the findings may be represented by a polynomial regression model with $p = 3$.

## 6.7.7 TRAINING, VALIDATION, AND TEST SETS

To prevent overfitting and evaluate the performance of a model, datasets are often split into training, validation, and testing sets [40]. Training the model using examples is how modeling is accomplished. The model is still insufficient to differentiate between training and test data and outcomes. Then the retraining of the model on the validation set is done, which allows for more refinement. Figure 6.4 shows that providing the model can withstand more complex input data, increasing the level of detail (a higher $p$) often results in better interpolation, at the expense of a reduction in its ability to forecast the training set. The optimal distribution of difficulty is found when the validation set contains as few solutions as possible [45]. Finally, the test set employs a subset of data points that are new to the model with the aim of maximizing sophistication.

### 6.7.8 CROSS-VALIDATION

Leaving so much data unobserved during model training is problematic in real-world circumstances when dataset sizes are small. Validation procedures are one solution to this problem. Cross-validation (CV) is useful for adjusting hyperparameters and choosing the right model. CV is a mathematical technique for estimating the findings' applicability [28]. The holdout between the training and test sets may be used as a one-way validation approach for any dataset [25,29,30]. As can be seen in Figure 6.5a, the holdout approach [28] works well when dealing with such massive data. Despite this, there are still benefits to using three-way validation. More practice data may be obtained with the three-way holdout approach [26,27,48]. A uniform frequency distribution throughout the data ensures that it was all collected from the same population. Quantifications of data quantities are quite usual [25–27].

The $k$-fold cross-validation approach ($k$-CV) (data partitioning into $k$ subsets and testing each against each other, where we rely on $k$-CV) may help with this issue, despite its seeming difficulty [43]. Similarly, $k$-fold cross-validation employs a single criterion, $k$, to assess each sample set. The term "$k$-fold validation" is used to describe this procedure. Leave-one-out cross-validation with a total of $k$ samples and $k$-fold forward cross-validation ($k$-CV) are two examples of common $k$-fold cross-validation procedures, as shown in Figure 6.5b and c. It divides the total number of training iterations into $k$ chunks, each of which carries out training on $k-1$ of the folds and validation on the remaining folds. Until the end of the $k$ validation sets, it is repeated. To determine the overall accuracy of the model, we simply average its performance over all $k$ validation folds.

### 6.7.9 METHODS OF REGULARIZATION

Regularization techniques [45] like least absolute shrinkage and selection operator [50], Ridge [51], and Elastic Net [45,52] may also be used to reduce the complexity of a model by filtering out irrelevant modeling terms. The basic idea of regularization is a costing characteristic, which consists of (a) the model's capacity to predict known data and (b) a supplemental notion that imposes a penalty on complex models. Because of this decrease in feature cost, unimportant words (those that do not significantly improve the model's accuracy) are eliminated. Changing the punitive weight allows one to fine-tune the model's complexity and provide the most accurate prediction possible within the validity range [53].

### 6.7.10 UPSKILLING THE MANPOWER

Large and quick improvements in data utilization can be achieved by better equipping our employees. We have a lot of smart people in our field, both technically

**(a)**

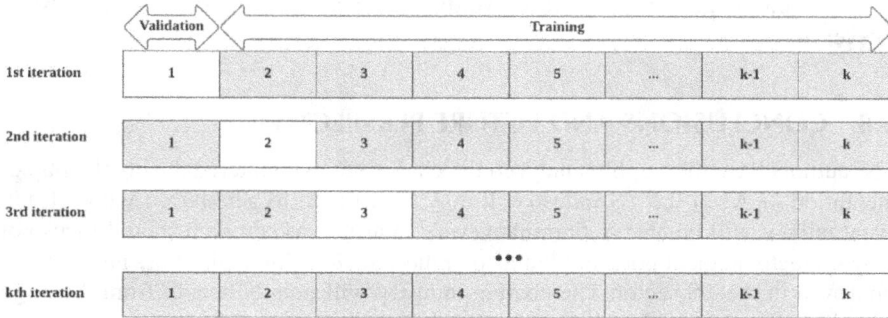

**(b)**

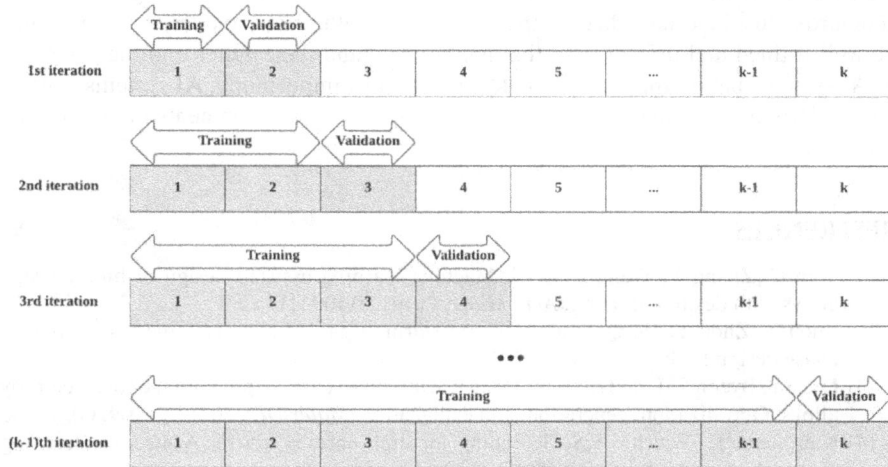

**(c)**

**FIGURE 6.5** Measurement strategies for ML (a) holdout method, (b) leave-one-out cross-validation for total $k$ samples, and (c) $k$-fold FCV [49]. Permissions from Elsevier.

and in terms of running successful businesses. Making data useful requires training our current workforce and using these data approaches and technologies to supplement or aid our current leaders and specialists. Due in part to the novelty of these talents and in part to the quick hiring of experienced practitioners in these skills at high pay by tech giants, our sector is experiencing a shortage of fresh talent in AI, ML, and big data. In time, the scarcity of fresh talent will be addressed by the boom in new university courses as well as digital and data apprenticeships. Tools that clean, analyze, and model data without the requirement for familiarity with code or data have proliferated in recent years. This development and democratization of data tools are comparable to the introduction of Windows in the early '90s. Windows made it so that even those who didn't know anything about computers could use them effectively. The new breed of data tools performs the same function. Alteryx, RapidMiner, and Microsoft Azure ML are just a few examples of data technologies that aim to make data science and AI/ML more accessible to everyone.

## 6.8   CONCLUSIONS AND FUTURE PERSPECTIVE

The authors have thoroughly analyzed the challenges encountered during the implementation of AI in the TS industry. It may be difficult to get started with AI, but eventually, it will be able to dominate every industry. AI research should focus not only on technological potential but also on how such technologies may be used and improved in the TS sector. The coating industry will greatly benefit from the large knowledge base that AI will provide, and the whole coating deposition process will be infused with a high degree of intelligence. Over time, and without a lot of human interaction, AI will become an essential element of the TS business. Regulations and standards must be established by the appropriate authorities, and the user base must be well-trained and informed so that they understand their duties and the functions of AI systems before they can be widely used. Most importantly, AI systems need to be regularly updated and integrated into the day-to-day developments in the thermal spray industry.

## REFERENCES

[1] Wen C, Zhang Y, Wang C, et al. Machine learning assisted design of high entropy alloys with desired property. *Acta Mater.* 2019;170:109–117.
[2] Zhou Z, Zhou Y, He Q, et al. Machine learning guided appraisal and exploration of phase design for high entropy alloys. *NPJ Comput Mater.* 2019;5:128.
[3] Lee SY, Byeon S, Kim HS, et al. Deep learning-based phase prediction of high-entropy alloys: Optimization, generation, and explanation. *Mater Des.* 2021;197:109260.
[4] Kaufmann K, Vecchio KS. Searching for high entropy alloys: A machine learning approach. *Acta Mater.* 2020;198:178–222.
[5] Li J, Xie B, Fang Q, et al. High-throughput simulation combined machine learning search for optimum elemental composition in medium entropy alloy. *J Mater Sci Technol.* 2021;68:70–75.
[6] Chattopadhyay C, Prasad A, Murty BS. Phase prediction in high entropy alloys-A kinetic approach. *Acta Mater.* 2018;153:214–225.
[7] Kamnis S, Sfikas AK, Gonzalez S, et al. A new cooling-rate-dependent machine learning feature for the design of thermally sprayed high-entropy alloys. *J Thermal Spray Technol.* 2022;32:401–414.
[8] Oliynyk AO, Antono E, Sparks TD, et al. High-throughput machine-learning-driven synthesis of full-Heusler compounds. *Chem Mater.* 2016;28:7324–7331.
[9] Challener C. *Leveraging Big Data, Artificial Intelligence, and Machine Learning in the Coatings Industry.* CoatingsTech; 2019;16(09), September 2019. Available from: https://www.paint.org/coatingstech-magazine/articles/leveraging-big-data-artificial-intelligence-and-machine-learning-in-the-coatings-industry/

[10] Cai Z, Liang H, Quan S, et al. Computer-aided robot trajectory auto-generation strategy in thermal spraying. *J Thermal Spray Technol.* 2015;24:1235–1245.

[11] Ding S, He P, Ma G, et al. Numerical simulation and experimental study of heat accumulation in cylinder parts during internal rotating plasma spraying. *J Thermal Spray Technol.* 2019;28:1636–1650.

[12] Gu S, Kamnis S. Numerical modelling of in-flight particle dynamics of non-spherical powder. *Surf Coat Technol.* 2009;203:3485–3490.

[13] Kout A, Wiederkehr T, Müller H. Efficient stochastic simulation of thermal spray processes. *Surf Coat Technol.* 2009;203:1580–1595.

[14] Trifa F-I, Montavon G, Coddet C. Model-based expert system for design and simulation of APS coatings. *J Thermal Spray Technol.* 2007;16:128–139.

[15] Barradas S, Guipont V, Molins R, et al. Laser shock flier impact simulation of particle-substrate interactions in cold spray. *J Thermal Spray Technol.* 2007;16:548–556.

[16] Zhou H, Li C, Bennett C, et al. Numerical analysis of deformation behavior and interface bonding of Ti6Al4V particle after subsequent impact during cold spraying. *J Thermal Spray Technol.* 2021;30:1093–1106.

[17] Dou Y, Luo J, Qi L, et al. Generation mechanism and suppression method of landing error of two successively deposited metal droplets caused by coalescence and solidification. *Int J Heat Mass Transf.* 2021;172:121100.

[18] Zheng YZ, Li Q, Zheng ZH, et al. Modeling the impact, flattening and solidification of a molten droplet on a solid substrate during plasma spraying. *Appl Surf Sci.* 2014;317:526–533.

[19] Fardan A, Berndt CC, Ahmed R. Numerical modelling of particle impact and residual stresses in cold sprayed coatings: A review. *Surf Coat Technol.* 2021;409:126835.

[20] Lee J, Park S-J, Oh Y-S, et al. Fragmentation behavior of Y2O3 suspension in axially fed suspension plasma spray. *Surf Coat Technol.* 2017;309:456–461.

[21] Oberste Berghaus J, Marple BR. High-velocity oxy-fuel (HVOF) suspension spraying of mullite coatings. *J Thermal Spray Technol.* 2008;17:671–678.

[22] Akbarnozari A, Amiri S, Dolatabadi A, et al. Analysis of scattering light from in-flight particles in suspension plasma spray for size measurement. *J Thermal Spray Technol.* 2019;28:678–689.

[23] Liu M, Zhang Y, Dong W, et al. Grey modeling for thermal spray processing parameter analysis. *Grey Syst Theory Appl.* 2020; 10(3): 265–279.

[24] Olleak A, Xi Z. Calibration and validation framework for selective laser melting process based on multi-fidelity models and limited experiment data. *J Mech Des.* 2020;142:81701.

[25] Frochte J. *Maschinelles Lernen: Grundlagen und Algorithmen in Python.* Munich: Carl Hanser Verlag GmbH Co KG; 2020.

[26] Kang S, Iwana BK, Uchida S. Complex image processing with less data-Document image binarization by integrating multiple pre-trained U-Net modules. *Pattern Recognit.* 2021;109: p. 107577.

[27] Ferreyra-Ramirez A, Aviles-Cruz C, Rodriguez-Martinez E, et al. An improved convolutional neural network architecture for image classification. In: Carrasco JA, Martínez JF, Olvera JA, et al., editors. *Pattern Recognition, 11th Mex Conf MCPR 2019 Querétaro,* Mex June 26–29, 2019 Proc. Lecture No. Cham: Springer Nature Switzerland AG; 2020. pp. 89–104.

[28] Raschka S. Model evaluation, model selection, and algorithm selection in machine learning. arXiv. 2018.

[29] Lecun Y, Bengio Y, Hinton G. Deep learning. *Nature.* 2015;521:436–444.

[30] Voulodimos A, Doulamis N, Doulamis A, et al. Deep learning for computer vision: A brief review. *Comput Intell Neurosci.* 2018; 2018: p. 7068349.

[31] Brownlee J. *Machine Learning Mastery With Python: Understand Your Data, Create Accurate Models, and Work Projects End-to-End. v1.20*. Victoria: Machine Learning Mastery; 2021.

[32] Vyas S, Jain SS, Choudhary I, et al. Study on use of AI and big data for commercial system. *Proc -2019 Amity Int Conf Artif Intell AICAI 2019*. 2019. pp. 737–739.

[33] Rusu O, Halcu I, Grigoriu O, et al. Converting unstructured and semi-structured data into knowledge. *Proc - RoEduNet IEEE Int Conf*. 2013.

[34] Bulgarevich DS, Tsukamoto S, Kasuya T, et al. Pattern recognition with machine learning on optical microscopy images of typical metallurgical microstructures. *Sci Rep* [Internet]. 2018;8:3–9. Available from: https://doi.org/10.1038/s41598-018-20438-6.

[35] Kuhn M, Johnson K. Over-Fitting and Model Tuning. In: Kuhn M, Johnson K. editors, *Applied Predictive Modeling*. New York: Springer 2013; pp. 61-92.

[36] Hancer E, Xue B, Zhang M. A survey on feature selection approaches for clustering. *Artif Intell Rev* [Internet]. 2020;53:4519–4545. Available from: https://doi.org/10.1007/s10462-019-09800-w.

[37] Jordan MI, Mitchell TM. Machine learning: Trends, perspectives, and prospects. *Science*. 2015;349:255–260.

[38] Efromovich S. *The Multivarient Normal Distribution*. Ney York: Springer Series in Statistics, Springer-Verlag; 1999.

[39] Liu Y, Zhao T, Ju W, et al. Materials discovery and design using machine learning. *J Mater* [Internet]. 2017;3:159–177. Available from: https://doi.org/10.1016/j.jmat.2017.08.002.

[40] Shobha G, Rangaswamy S. Machine learning. In: Gudivada VN, Rao CR. editors, *Handbook of Statistics*. 1st ed. Amsterdam: Elsevier B.V.; 2018; 38: 197-228. Available from: https://doi.org/10.1016/bs.host.2018.07.004.

[41] Willmott CJ, Matsuura K. Advantages of the mean absolute error (MAE) over the root mean square error (RMSE) in assessing average model performance. *Clim Res*. 2005;30:79–82.

[42] Hagan MT, Menhaj M. Training feedforward networks with the Marquardt algorithm. *IEEE Trans Neural Netw*. 1994;5:989–993.

[43] Alcobaça E, Mastelini SM, Botari T, et al. Explainable machine learning algorithms for predicting glass transition temperatures. *Acta Mater*. 2020;188:92–100.

[44] Lee LC, Liong CY, Jemain AA. Validity of the best practice in splitting data for hold-out validation strategy as performed on the ink strokes in the context of forensic science. *Microchem J* [Internet]. 2018;139:125–133. Available from: https://doi.org/10.1016/j.microc.2018.02.009.

[45] Emmert-Streib F, Dehmer M. Evaluation of regression models: Model assessment, model selection and generalization error. *Mach Learn Knowl Extr*. 2019;1:521–551.

[46] Pedregosa F, Varoquaux G, Gramfort A, et al. Scikit-learn: Machine learning in python. *J Mach Learn Res*. 2011;12:2825–2830.

[47] Halilaj E, Rajagopal A, Fiterau M, et al. Machine learning in human movement biomechanics: Best practices, common pitfalls, and new opportunities. *J Biomech* [Internet]. 2018;81:1–11. Available from: https://doi.org/10.1016/j.jbiomech.2018.09.009.

[48] Chang M-W, Devlin J, Dragan A, et al. *Artificial Intelligence: 1. Introduction*. In 4th ed. Russell S, Norvig P, editors. Artificial Intelligence: A Modern Approach, Hoboken, NJ: Pearson Education, Inc.; 2020, pp. 1-35.

[49] Xiong Z, Cui Y, Liu Z, et al. Evaluating explorative prediction power of machine learning algorithms for materials discovery using k-fold forward cross-validation. *Comput Mater Sci* [Internet]. 2020;171:109203. Available from: https://doi.org/10.1016/j.commatsci.2019.109203.

[50] Algamal ZY. Shrinkage parameter selection via modified cross-validation approach for ridge regression model. *Commun Stat Simul Comput* [Internet]. 2020;49:1922–1930. Available from: https://doi.org/10.1080/03610918.2018.1508704.

[51] Wilcox RR. Multicolinearity and ridge regression: results on type I errors, power and heteroscedasticity. *J Appl Stat.* 2019;46:946–957.

[52] Li N, Yang H, Yang J. Nonnegative estimation and variable selection via adaptive elastic-net for high-dimensional data. *Commun Stat - Simul Comput* [Internet]. 2019;0:1–17. Available from: https://doi.org/10.1080/03610918.2019.1642484.

[53] Wang Y, Yang XG, Lu Y. Informative gene selection for microarray classification via adaptive elastic net with conditional mutual information. *Appl Math Model* [Internet]. 2019;71:286–297. Available from: https://doi.org/10.1016/j.apm.2019.01.044.

# 7 Neural Network Model for Wear Prediction of Coatings
## Case Study

## ABBREVIATIONS

| | |
|---|---|
| **AI** | Artificial intelligence |
| **ANN** | Artificial neural network |
| **CV** | Cross validation |
| **ML** | Machine learning |
| **MSE** | Mean square error |
| **PS** | Plasma spray |
| **R2** | Coefficient of determination |
| **RMSE** | Mean absolute error root mean square error |
| **TS** | Thermal spraying |

## 7.1 BRIEF INTRODUCTION TO WEAR AND ARTIFICIAL NEURAL NETWORK TECHNIQUE

Slurry erosion is a distinct form of wear of the hydro-mechanical component surface caused by the continuous impact of the slurry particles [1–3]. Kinetic energy is transferred from moving slurry particles to the solid surface, resulting in erosion. Failure of a material's surface might result from a directed impact or random collisions with solid particles [4]. Factors such as flow rate, duration, slurry concentration, impingement angle, material characteristics, and coating parameters are all key factors in erosion phenomena [5–7]. In this chapter, an artificial neural network (ANN) was used to predict the slurry erosion for Ni-20$Cr_2O_3$, i.e., a coating produced using the high-velocity oxy-fuel technique on stainless steel 316L. When traditional statistical and analytical approaches no longer work, a neural network might be a useful alternative. For ANN to make a forecast, it must first be trained or taught. Instead of using knowledge-based information, ANN employs a prediction approach [8]. Optimization of weights via an adaptive function that is knowledge-dependent is how ANN learns [9]. In this regard, feed-forward neural networks have become more popular in recent years [10]. However, the ANN model can be trusted to provide the lowest possible error rate in output [11,12]. Successful predictions of material wear behavior using ANN models have been presented by some researchers [13,14]. The methods of Neural Networks are discussed in detail in the first two chapters.

 DOI: 10.1201/9781003400660-7

In this section, the construction of an ANN model in the programming language MATLAB® will be learned through a case study. MATLAB's NNTOOL may be used to implement the ANN.

## 7.2　SAMPLE DATA

Data used for the ANN model is divided into input and output files. The input files contain different parameters in a matrix. For example, the $7 \times 36$ matrix was used as input data in the present case study. Table 7.1 shows the dataset used in the present study. This data set is converted row-wise for an accurate reading. However, the output file should contain the same rows as the input file for an accurate reading of the data. Against the $7 \times 36$ Matrix data, the output file used in the present study was $1 \times 36$ matrix. The output was the actual experimental data from erosion experiments.

## 7.3　SELECTION OF NNTOOL

There are different types of neural network tools used for different purposes, such as classification learning, clustering, curve fitting, pattern recognition, time series analysis, and regression learning. Figure 7.1 shows the various NN techniques available in the MATLAB tool. In the present case study, the NN fitting tool was selected for the prediction of coating and stainless steel 316L.

## TABLE 7.1
### Sample Data Used for ANN Modeling

| | Inputs | | | | | | | Output |
|---|---|---|---|---|---|---|---|---|
| Materials | Revolutions/ min (RPM) | Concentration | Time | Particle Diameter | Impact Angle | Porosity | Hardness | Wear (g/ m² min) |
| SS 316L | 600 | 30 | 180 | 196.5 | 0 | 0 | 212 | 0.31251 |
| | 900 | 30 | 180 | 196.5 | 0 | 0 | 212 | 0.41357 |
| | 1200 | 30 | 180 | 196.5 | 0 | 0 | 212 | 0.53394 |
| | 1500 | 30 | 180 | 196.5 | 0 | 0 | 212 | 0.67406 |
| | 1500 | 30 | 180 | 196.5 | 0 | 0 | 212 | 0.67406 |
| | 1500 | 40 | 180 | 196.5 | 0 | 0 | 212 | 0.88132 |
| | 1500 | 50 | 180 | 196.5 | 0 | 0 | 212 | 0.99503 |
| | 1500 | 60 | 180 | 196.5 | 0 | 0 | 212 | 1.16118 |
| | 1500 | 60 | 90 | 196.5 | 0 | 0 | 212 | 1.46321 |
| | 1500 | 60 | 120 | 196.5 | 0 | 0 | 212 | 1.27321 |
| | 1500 | 60 | 150 | 196.5 | 0 | 0 | 212 | 1.19901 |
| | 1500 | 60 | 180 | 45.6 | 0 | 0 | 212 | 0.22362 |
| | 1500 | 60 | 180 | 93.4 | 0 | 0 | 212 | 0.30353 |
| | 1500 | 60 | 180 | 121.7 | 0 | 0 | 212 | 0.34065 |
| | 1500 | 60 | 180 | 257.8 | 0 | 0 | 212 | 0.51076 |
| | 1500 | 30 | 180 | 196.5 | 30 | 0 | 212 | 0.91736 |
| | 1500 | 30 | 180 | 196.5 | 45 | 0 | 212 | 0.89052 |
| | 1500 | 30 | 180 | 196.5 | 60 | 0 | 212 | 0.71529 |

*(Continued)*

**TABLE 7.1 (*Continued*)**
**Sample Data Used for ANN Modeling**

| Materials | Revolutions/ min (RPM) | Concentration | Time | Particle Diameter | Impact Angle | Porosity | Hardness | Wear (g/ m² min) |
|---|---|---|---|---|---|---|---|---|
| | | | | **Inputs** | | | | **Output** |
| Ni-Cr₂O₃ | 600 | 30 | 180 | 196.5 | 0 | 1.13 | 316 | 0.19686 |
| coating | 900 | 30 | 180 | 196.5 | 0 | 1.13 | 316 | 0.31006 |
| | 1200 | 30 | 180 | 196.5 | 0 | 1.13 | 316 | 0.3979 |
| | 1500 | 30 | 180 | 196.5 | 0 | 1.13 | 316 | 0.45498 |
| | 1500 | 30 | 180 | 196.5 | 0 | 1.13 | 316 | 0.45498 |
| | 1500 | 40 | 180 | 196.5 | 0 | 1.13 | 316 | 0.67574 |
| | 1500 | 50 | 180 | 196.5 | 0 | 1.13 | 316 | 0.84035 |
| | 1500 | 60 | 180 | 196.5 | 0 | 1.13 | 316 | 1.05892 |
| | 1500 | 60 | 90 | 196.5 | 0 | 1.13 | 316 | 1.22021 |
| | 1500 | 60 | 120 | 196.5 | 0 | 1.13 | 316 | 1.11174 |
| | 1500 | 60 | 150 | 196.5 | 0 | 1.13 | 316 | 1.06466 |
| | 1500 | 60 | 180 | 45.6 | 0 | 1.13 | 316 | 0.00604 |
| | 1500 | 60 | 180 | 93.4 | 0 | 1.13 | 316 | 0.00795 |
| | 1500 | 60 | 180 | 121.7 | 0 | 1.13 | 316 | 0.01071 |
| | 1500 | 60 | 180 | 257.8 | 0 | 1.13 | 316 | 0.01495 |
| | 1500 | 30 | 180 | 196.5 | 30 | 1.13 | 316 | 0.64828 |
| | 1500 | 30 | 180 | 196.5 | 45 | 1.13 | 316 | 0.65412 |
| | 1500 | 30 | 180 | 196.5 | 60 | 1.13 | 316 | 0.66001 |

*Source:* PhD Thesis; Singh [15].
ANN, artificial neural network.

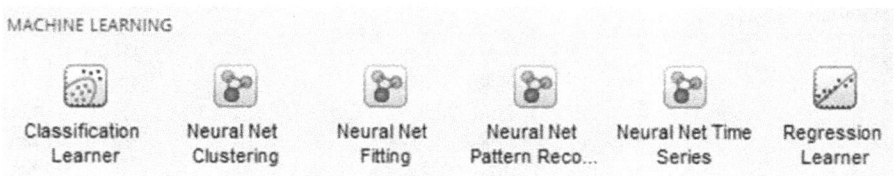

FIGURE 7.1 Various artificial neural network techniques.

## 7.4 DEVELOPMENT OF ANN ARCHITECTURE

After selecting the NN fitting tool, a window will display, as shown in Figure 7.2. This window displays the basic structure of the NN model. After clicking next, the next window shows the option to select the input and output datasets, as shown in Figure 7.3. The next step in the development of the NN model is the selection of training, validation, and testing data in percentages, as shown in Figure 7.4. Afterward, suitable hidden neurons and training algorithms are selected (Figures 7.5 and 7.6). These are the basics of how the NN fitting app is used to develop the NN model for wear prediction.

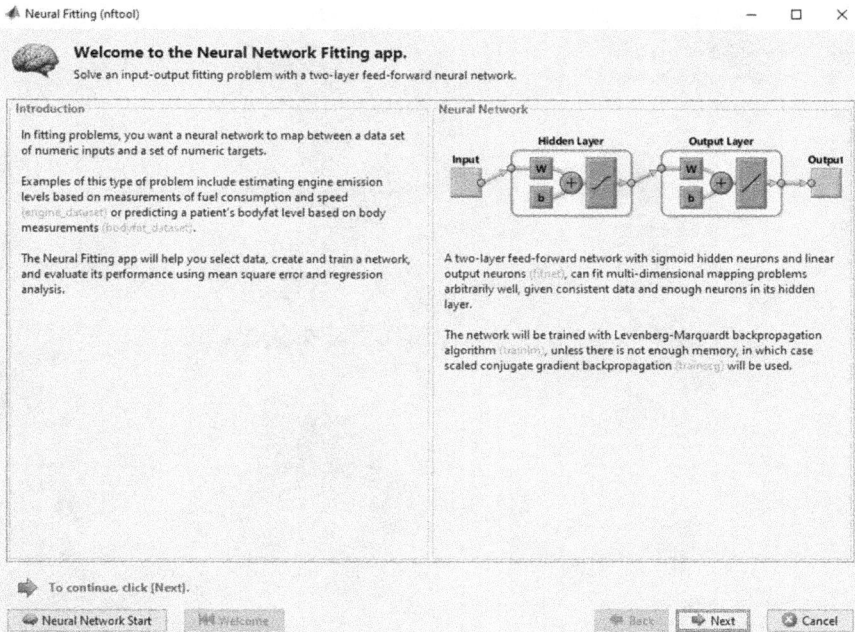

**FIGURE 7.2** Main window of the NN tool.

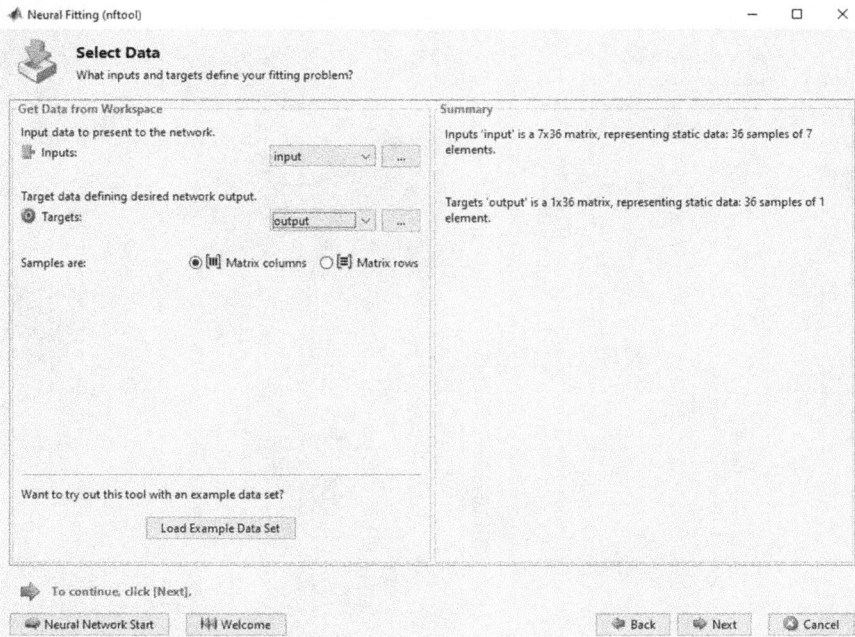

**FIGURE 7.3** Input and output selection windows of the NN tool.

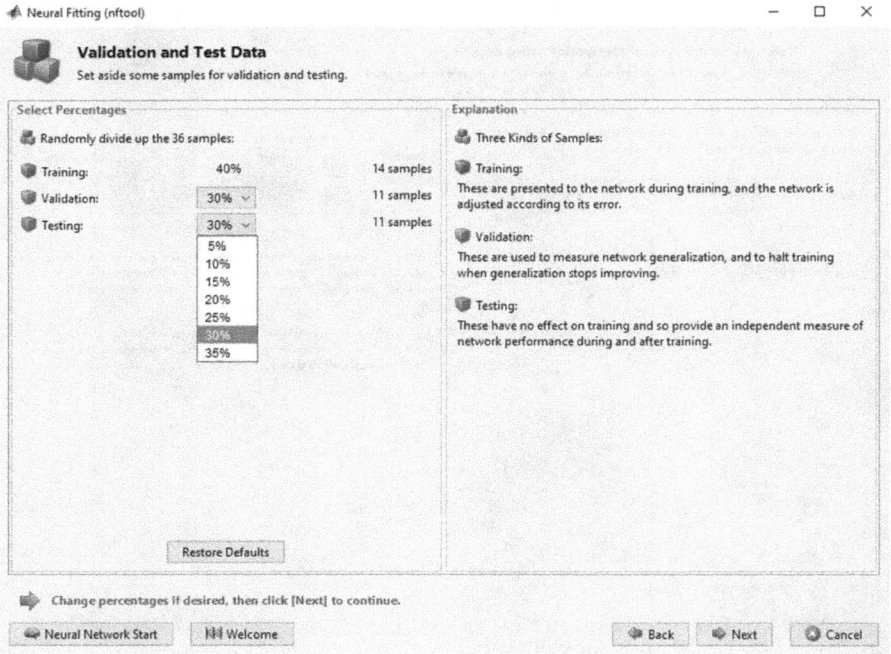

**FIGURE 7.4**   Selection of training, validation, and testing data in the NN tool.

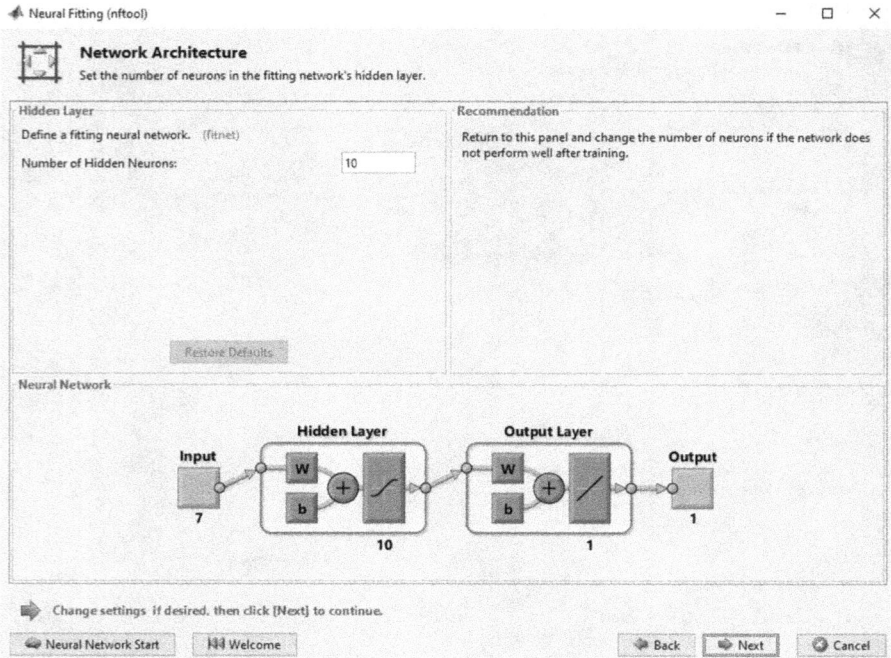

**FIGURE 7.5**   Selection of hidden neurons in the NN tool.

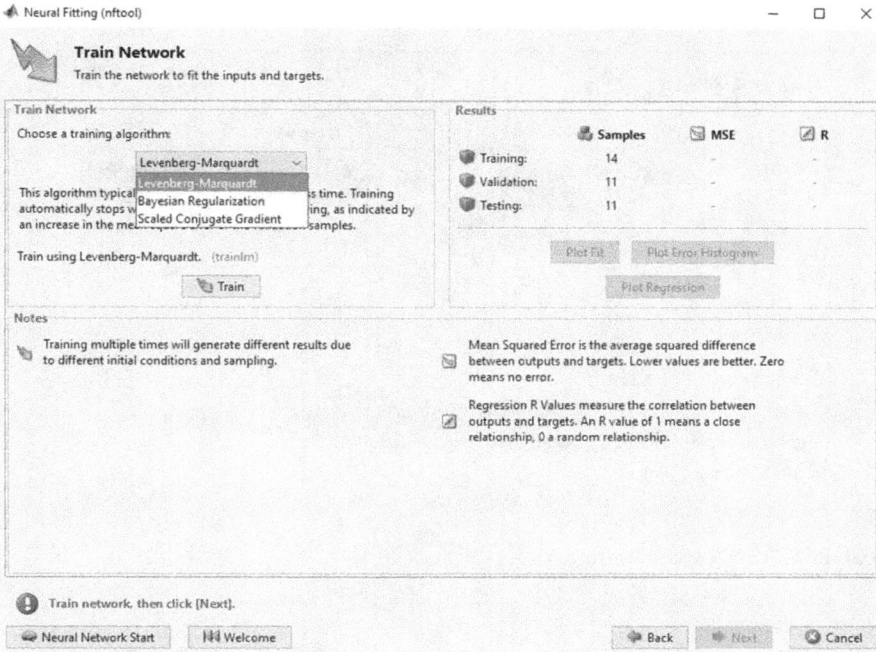

**FIGURE 7.6**   Selection of training algorithms in the NN tool.

## 7.5   ANN ARCHITECTURE

In this particular research endeavor, the adaptive learning function known as Gradient Descent with Momentum (abbreviated as learngdm) was utilized [16]. An ANN model was trained using the MATLAB NNtool with input parameters and an output parameter, i.e., erosion rate. For this analysis, the numerous hidden neurons were numbered as 9, 10, 11, 12, 15, 18, 20, 21, 24, 30, and 35. The 12 hidden neurons predicted the best results in erosion rate in the present study. The present NN model's architecture is illustrated in Figure 7.7. The Levenberg-Marquardt algorithms were utilized for the learning of the ANN, which provides the fastest convergence of the model [17,18]. The training and testing functions were standardized in the range of 0.1–0.9 with the help of the equation written below:

$$y = 0.1 + 0.8\left(\frac{x - x_{\min}}{x_{\max} - x_{\min}}\right) \tag{7.1}$$

## 7.6   TRAINING AND VALIDATION OF THE ANN MODEL

In this study, the training, validation, and testing data were selected at 75%, 15%, and 15%, respectively. The network was trained after 17 epochs. The NN model showed the best performance at the first epoch after running the epochs several times, as shown in Figure 7.8.

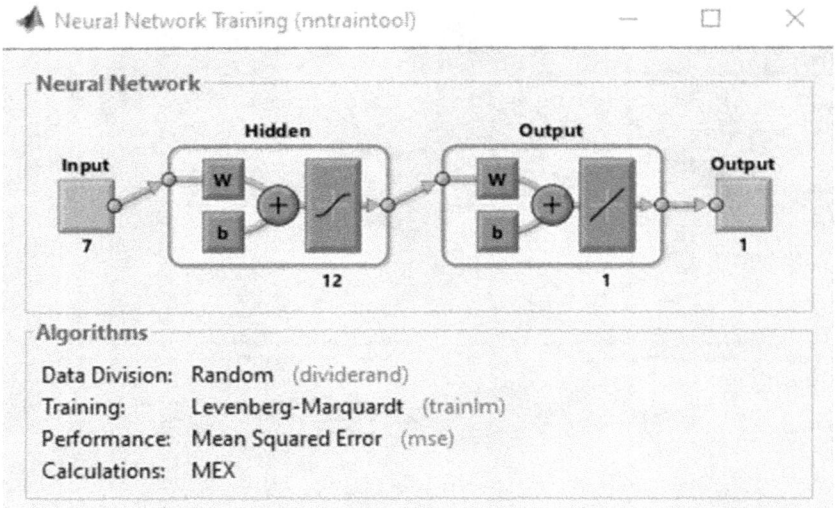

**FIGURE 7.7**  NN architecture.

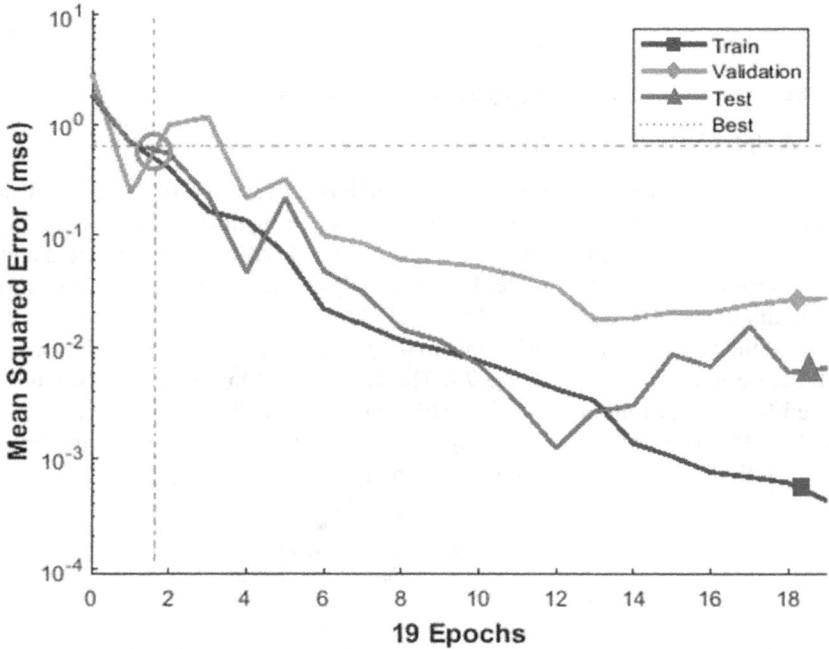

**FIGURE 7.8**  Best performance of the NN model.

## 7.7 WEAR RESULTS FROM THE ANN MODEL

The values that are predicted by the model have a high degree of correspondence with the values that are discovered in real experiments. The error histogram that was generated by the ANN when it was being trained can be found in Figure 7.9. The performance of the ANN model is evaluated based on an error percentage that falls between 0% and 6%. Figure 7.10 displays the Pearson coefficient ($R$) for the training, validation, and testing phases. The Pearson correlation coefficient ($R$) [19–22] is given by:

$$R = \frac{A \sum pq - \left( \sum p \sum q \right)}{\sqrt{\left[ A \sum p^2 - \left( \sum p \right)^2 \right]\left[ A \sum q^2 - \left( \sum q \right)^2 \right]}} \qquad (7.2)$$

where $A$ is the pair scores. $\sum pq$ is the product of paired scores. Symbols $\sum p$ and $\sum p$ are the SS (sum of scores). The values of the Pearson coefficient calculated by ANN were $9.92003 \times 10^{-1}$, $9.63304 \times 10^{-1}$, and $9.87754 \times 10^{-1}$ for training, validation, and testing, respectively.

Root Mean Square Error (RMSE) [22] was also calculated as a performance measure:

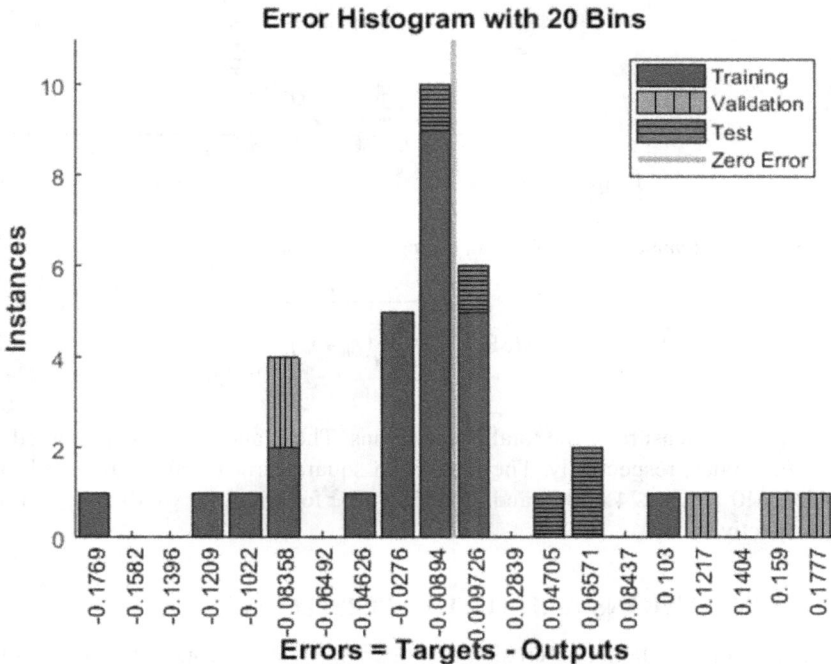

**FIGURE 7.9** Error histogram of the NN model.

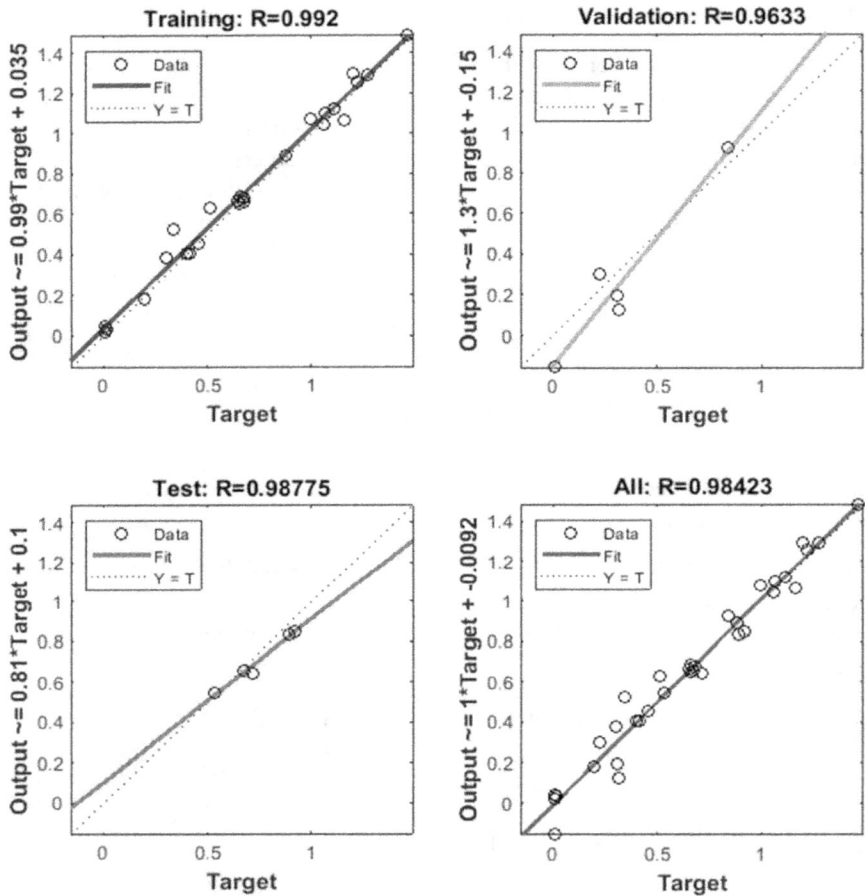

FIGURE 7.10   *R*-value of the NN model during different phases.

$$\text{RMSE} = \sqrt{\frac{1}{x}\sum_{n=1}^{k}(r_n - \overline{r}_n)^2} \qquad (7.3)$$

where $x$ is the measure of the total observations. The $r_n$ and $\overline{r}_n$ are the predicted and observed values, respectively. The root mean square error (MSE) was found to be $3.34533 \times 10^{-3}$, $1.76224 \times 10^{-2}$, and $2.65858 \times 10^{-3}$ for training, validation, and testing, respectively.

## 7.8   CONCLUSIONS AND FUTURE PERSPECTIVE

Neural network prediction of erosion rate in stainless steel 316L and Ni-based high-velocity oxy-fuel coating, i.e., Ni-20Cr$_2$O$_3$ was carried out in this chapter. The results of this case study indicate that the newly developed ANN model has improved prediction ability. This error range of 0–6% demonstrates the effectiveness of the ANN model. The Pearson coefficient and MSE readings show that the ANN model prediction error was very low. The ANN tools are very immersive in terms of prediction of erosion wear. In the future, the training data can be supplied less frequently than the data supplied in this case study, and more data can be predicted by supplying the less frequently supplied data.

## REFERENCES

[1] Bitter JGA. A study of erosion phenomena part I. *Wear.* 1963;6:169–190.

[2] Finne I. Erosion of surfaces. *Wear.* 1960;3:87–103.

[3] Finnie I. Some observations on the erosion of ductile metals. *Wear.* 1972;19:81–90.

[4] Gupta R, Singh SN, Seshadri V. Study on the uneven wear rate in a slurry pipeline on the basis of mesurements in a pot tester. *Wear.* 1995;184:169–178.

[5] Gandhi BK, Singh SN, Seshadri V. Study of the parametric dependence of erosion wear for the parallel flow of solid-liquid mixtures. *Tribol Int.* 1999;32:275–282.

[6] Vieira RE, Mansouri A, McLaury BS, et al. Experimental and computational study of erosion in elbows due to sand particles in air flow. *Powder Technol* [Internet]. 2016;288:339–353. Available from: https://doi.org/10.1016/j.powtec.2015.11.028.

[7] Vuorinen E, Ojala N, Heino V, et al. Erosive and abrasive wear performance of carbide free bainitic steels - Comparison of field and laboratory experiments. *Tribol Int* [Internet]. 2016;98:108–115. Available from: https://doi.org/10.1016/j.triboint.2016.02.015.

[8] Mihalakakou G, Santamouris M, Asimakopoulos D. The total solar radiation time series simulation in Athens, using neural networks. *Theor Appl Climatol.* 2000;66:185–197.

[9] Saleh B, Aly AA. Artificial neural network model for evaluation the effect of surface properties amendment on slurry erosion behavior of AISI 5117 steel. *Ind Lubr Tribol.* 2016;68:676–682.

[10] Mehrotra K, Mohan CK, Ranka S. *Elements of Artificial Neural Networks.* Cambridge, Massachusetts: MIT Press; 1997.

[11] Shuvho MBA, Chowdhury MA, Ahmed S, et al. Prediction of solar irradiation and performance evaluation of grid connected solar 80KWp PV plant in Bangladesh. *Energy Rep* [Internet]. 2019;5:714–722. Available from: https://doi.org/10.1016/j.egyr.2019.06.011.

[12] Shuvho BA, Chowdhury MA, Debnath UK. Analysis of artificial neural network for predicting erosive wear of nylon-12 polymer. *Mater Perform Charact.* 2019;8(1):288–300.

[13] Yetim AF, Codur MY, Yazici M. Using of artificial neural network for the prediction of tribological properties of plasma nitrided 316L stainless steel. *Mater Lett* [Internet]. 2015;158:170–173. Available from: https://doi.org/10.1016/j.matlet.2015.06.015.

[14] Sahu SPR, Satapathy A, Mishra D, et al. Tribo-performance analysis of fly ash-aluminum coatings using experimental design and ANN. *Tribol Trans.* 2010;53:533–542.

[15] Singh J. *Investigation on Slurry Erosion of Different Pumping Materials and Coatings.* Patiala, India: Thapar Institute of Engineering and Technology; 2019.

[16] Qian N. On the momentum term in gradient descent learning algorithms. *Neural Netw.* 1999;12:145–151.

[17] Hagan MT, Menhaj M. Training feedforward networks with the Marquardt algorithm. *IEEE Trans Neural Netw.* 1994;5:989–993.

[18] Levenberg K. A method for the solution of certain non-linear problems in least squares. *Q Appl Math.* 1944;2:164–168.

[19] Singh J. Tribo-performance analysis of HVOF sprayed 86WC-10Co4Cr & Ni-Cr2O3 on AISI 316L steel using DOE-ANN methodology. *Ind Lubr Tribol.* 2021;73:727–735.

[20] Benesty J, Chen J, Huang Y, et al. *Pearson Correlation Coefficient. Noise Reduct Speech Process Springer Top Signal Process*, Vol. 2. Berlin, Heidelberg: Springer-Verlag; 2009. pp. 37–40.

[21] Kumar P, Singh J, Singh S. Neural network supported flow characteristics analysis of heavy sour crude oil emulsified by ecofriendly bio-surfactant utilized as a replacement of sweet crude oil. *Chem Eng J Adv* [Internet]. 2022;11:100342. Available from: https://doi.org/10.1016/j.ceja.2022.100342.

[22] Singh J, Singh S. Neural network supported study on erosive wear performance analysis of Y2O3/WC-10Co4Cr HVOF coating. *J King Saud Univ - Eng Sci* [Internet]. 2022, In press. Available from: https://doi.org/10.1016/j.jksues.2021.12.005.

# 8 Implementation of Regression Models for Wear Analysis of Coating
## Case Study

## ABBREVIATIONS

| | |
|---|---|
| **AI** | Artificial intelligence |
| **ANN** | Artificial neural network |
| **C88** | Colmonoy 88 |
| **CA** | Classification accuracy |
| **EnBoost** | Ensemble Boost |
| **GPR** | Gaussian process regression |
| **LR** | Linear regression |
| **PS** | Plasma spray |
| **RMSE** | Root mean square error |
| **SVM** | Support vector machine |

## 8.1 BRIEF INTRODUCTION TO REGRESSION TECHNIQUES

Regression analysis [1] can assess the impact of inputs on outcomes. Figure 8.1 provides a visual representation of how regression models, e.g., linear [2], polynomial [3], or nonlinear functions [4]. Parametric regression uses the input variables to determine an output value, whereas nonparametric regression uses a kernel function to estimate values at neighboring sites [5]. The $k$-nearest neighbors method [6,7] is an additional way of representing a nonparametric regression. The key idea behind the $k$-nearest neighbor method is to estimate the output at a given input position by using the $k$ value, which is determined by averaging the outputs at neighboring positions. A common Gaussian process regression (GPR) method [8] calculates, for each input location, a probability distribution of the Gaussian type based on the multi-variant normal connection between the input and all other established points [9]—where the degree of correlation decreases with increasing distance. Significant advantages in terms of projected performance value volatility may be gained by using the GPR method, which is crucial for establishing the accuracy of the forecasts [10].

Parametric regression, in contrast to nonparametric regression, focuses on describing and minimizing an objective function (cost) utilizing an explicit analytical formulation involving inputs and outputs of a concerned problem [1]. The components of this formulation have been modified according to familiar points. The use

DOI: 10.1201/9781003400660-8

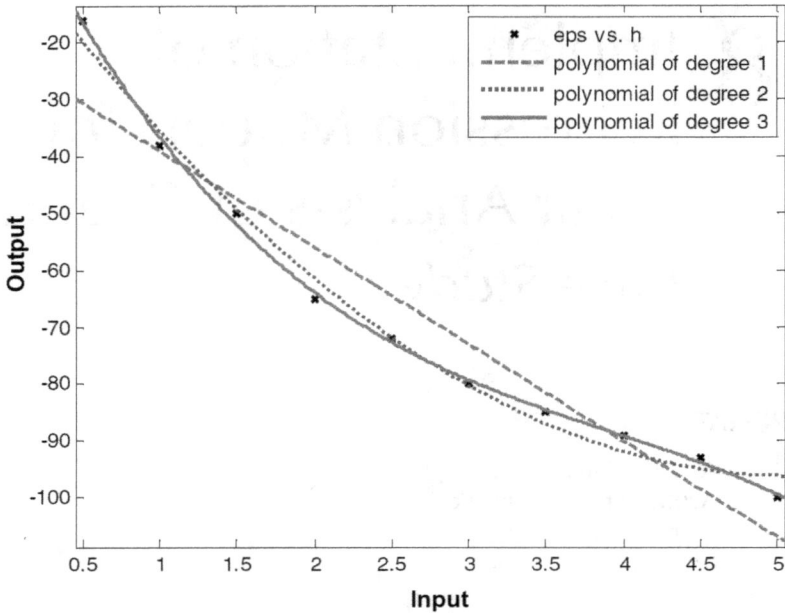

**FIGURE 8.1** Polynomial regression model (eps vs. $h$) example, where the degree ($p$) of the polynomial model is represented by the thickness of the lines shown as dots (green line, $p = 1$, blue line, $p = 2$, and red line, $p = 1$) [19]. Permssions under CC-BY-NC-ND 4.0.

of advanced ML algorithms for classification and regression should not be forgotten. Except for kernel-dependent functionality [11], many popular ML methods—such as artificial neural networks [12], support vector machines (SVMs) [13], random forests [14], and gradient boosting [15]—are based on convoluted nonlinear parametric formulae. Although these models have been shown to be useful for interpolating data [15], their interpretation is often poor due to their convoluted parametric formula organization [12] and limited hypothesizing ability [16]. Researching models and developing forecasts often entails two stages: (i) setup (which includes configuration and validation) and (ii) analysis. To make reliable predictions throughout the fitting or learning process, it is necessary to adjust the model's complexity (e.g., the maximum degree of polynomial regression) [17,18]. Below is a detailed description of the steps to minimize the complexity of the model.

## 8.2 CLASSIFICATION TECHNIQUES

Classification is an example of regression [20]. Classification issues differ from regression in that the output value is not fixed; instead, the model is trained using both the input characteristics and the target classes. There are just two groups in a binary categorization system. Either Class 1 or Class 2 output is possible. The constructed model after training can categorize characteristics into the classes in which it was taught. For optimal group separation in the input space, we may define it as the hyperground [20,21]. Figure 8.2 depicts a classifier for the linear and linear

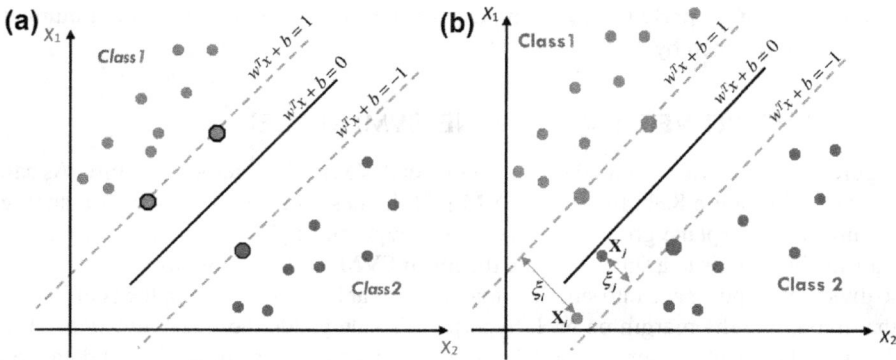

**FIGURE 8.2**    Maximum margins from each class of linearly separable data are shown (a) as support vectors (bold points), and (b) slack variables are established to minimize the error of misclassification for data that is not linearly separable [21]. Permissions from Elsevier.

non-separable cases. Figure 8.2a depicts a linear hyperplane for a given set of training data $(x_i)$, which is represented by straight lines.

$$w^T + b = 0 \tag{8.1}$$

In the given equation, the value of symbol $i$ may be 1, 2, 3, ..., $n$. The symbols $w$ and $b$ above stand for the $n$-dimensional vector and bias term, respectively, in the aforementioned equation. For the linearly separable situation, the hyperplane takes into consideration two requirements: (a) the amount of error in the separation data should be as low as possible, and (b) the distances between the clustered (class) data should be as large as possible [21]. In this example, we distinguish between two groups based on their location relative to the hyperplane's border, the left $(y = 1)$ and right $(y = 1)$ sides. This is why the separation is managed as follows:

$$w^T + b \begin{cases} \geq 1\{y_i = 1\} \\ \leq -1\{y_i = -1\} \end{cases} \tag{8.2}$$

However, as shown in Figure 8.2b, the optimum classification is dependent on a penalty function that minimizes the distance $(\xi_i)$ between the poor classifier and the good one.

$$F(\xi) = \sum_{i-1}^{N} \xi_i \tag{8.3}$$

Classification accuracy (CA) is a metric used to measure how well a model performs in solving classification issues [22], and it is given by:

$$CA = \frac{S}{N} \tag{8.4}$$

The number of properly labeled samples, denoted by $S$, is equal to the total number of samples, denoted by $N$.

## 8.3   SUPPORT VECTOR MACHINE (SVM) MODEL

Figure 8.3a depicts a simplified graphical model (SVM) of its internal structure. As can be shown in Figure 8.3b, the linear SVM [13] divides issues into classes by mapping features to hyperplane groups, and the kernel hyperplane [23–25] divides features by transferring details to a subclass. Like the linear SVM models, it employs straight lines to divide the input space into smaller, more manageable pieces. Finding the coefficients that maximize the margin on each hyperplane's side yields the linear functions [13]. The used kernel characterizes the relationship between training data and test data (i.e., the known class). Here, the equation generates nonlinear hyperplanes for categorization so that we may make the most efficient use of available space [23–25].

## 8.4   DECISION TREE MODELS

Decision trees are often implemented as a collection of smaller trees working as an ensemble using random forests [14]. Each constructed tree represents a choice (prediction), and the one with the most support is chosen as the model's forecast. The strength of a decision tree comes from its ensemble, a collection of prediction models that can outperform a single model operating alone. The larger the data collection, the taller the tree (or nodes) will be. By minimizing the prediction error on the validation set, underfitting and overfitting may be avoided if the appropriate requirements are used.

## 8.5   DATASET FOR REGRESSION MODELING

Data used for the regression modeling is divided into input and output files. The input files contain different parameters in a matrix. For example, the $9 \times 30$ matrix was used as input data in the present case study. Table 8.1 shows the dataset used in the

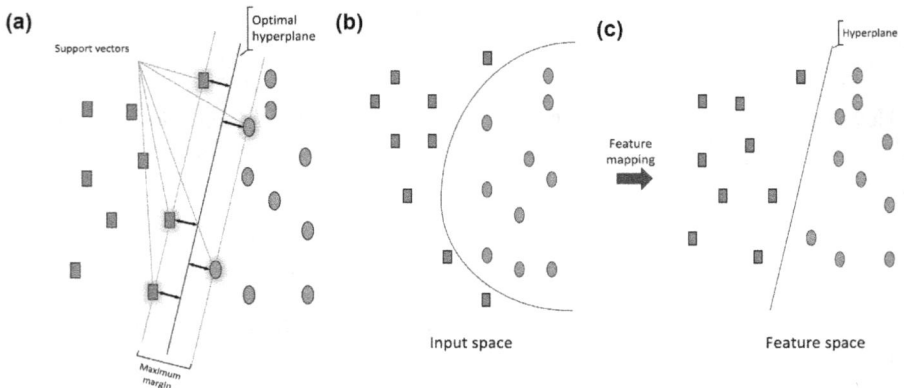

**FIGURE 8.3**   (a) Basic structure of a support vector machine (SVM), and (b, c) SVM classifications [27]. Permissions from Elsevier.

**TABLE 8.1**

**Sample Data Used for Regression Modeling**

| S. No. | V | C | k | CF | i. T | ii. $d_p$ | iii. A | iv. H | v. P | Materials | Output vi. E |
|---|---|---|---|---|---|---|---|---|---|---|---|
| 1 | 4.59 | 60 | 2.667 | 0.64 | 180 | 195.6 | 0 | 196 | 0 | SS 316L | 7.55E-05 |
| 2 | 4.59 | 60 | 2.667 | 0.64 | 180 | 195.6 | 30 | 196 | 0 | SS 316L | 5.65E-05 |
| 3 | 4.59 | 60 | 2.667 | 0.64 | 180 | 195.6 | 45 | 196 | 0 | SS 316L | 5.38E-05 |
| 4 | 4.59 | 60 | 2.667 | 0.64 | 180 | 195.6 | 60 | 196 | 0 | SS 316L | 3.72E-05 |
| 5 | 4.59 | 60 | 2.667 | 0.64 | 180 | 195.6 | 0 | 601 | 1.29 | C88 | 2.11E-05 |
| 6 | 4.59 | 60 | 2.667 | 0.64 | 180 | 195.6 | 30 | 601 | 1.29 | C88 | 2.15E-05 |
| 7 | 4.59 | 60 | 2.667 | 0.64 | 180 | 195.6 | 45 | 601 | 1.29 | C88 | 2.24E-05 |
| 8 | 4.59 | 60 | 2.667 | 0.64 | 180 | 195.6 | 60 | 601 | 1.29 | C88 | 2.36E-05 |
| 9 | 1.81 | 60 | 2.667 | 0.64 | 90 | 195.6 | 0 | 196 | 0 | SS 316L | 7.02E-06 |
| 10 | 2.71 | 60 | 2.667 | 0.64 | 90 | 195.6 | 0 | 196 | 0 | SS 316L | 1.05E-05 |
| 11 | 3.61 | 60 | 2.667 | 0.64 | 90 | 195.6 | 0 | 196 | 0 | SS 316L | 1.40E-05 |
| 12 | 4.59 | 60 | 2.667 | 0.64 | 90 | 195.6 | 0 | 196 | 0 | SS 316L | 1.78E-05 |
| 13 | 1.81 | 60 | 2.667 | 0.64 | 90 | 195.6 | 0 | 601 | 1.29 | C88 | 3.99E-06 |
| 14 | 2.71 | 60 | 2.667 | 0.64 | 90 | 195.6 | 0 | 601 | 1.29 | C88 | 5.99E-06 |
| 15 | 3.61 | 60 | 2.667 | 0.64 | 90 | 195.6 | 0 | 601 | 1.29 | C88 | 7.97E-06 |
| 16 | 4.59 | 60 | 2.667 | 0.64 | 90 | 195.6 | 0 | 601 | 1.29 | C88 | 1.01E-05 |
| 17 | 4.59 | 60 | 2.667 | 0.76 | 180 | 45.6 | 0 | 196 | 0 | SS 316L | 1.38E-05 |
| 18 | 4.59 | 60 | 2.667 | 0.65 | 180 | 93.4 | 0 | 196 | 0 | SS 316L | 2.45E-05 |
| 19 | 4.59 | 60 | 2.667 | 0.63 | 180 | 121.7 | 0 | 196 | 0 | SS 316L | 2.57E-05 |
| 20 | 4.59 | 60 | 2.667 | 0.53 | 180 | 257.8 | 0 | 196 | 0 | SS 316L | 3.76E-05 |

(*Continued*)

**TABLE 8.1 (Continued)**
**Sample Data Used for Regression Modeling**

| S. No. | Inputs | | | | | | | | | | Output |
| --- | --- | --- | --- | --- | --- | --- | --- | --- | --- | --- | --- |
| | V | C | k | CF | i. T | ii. $d_p$ | iii. A | iv. H | v. P | Materials | vi. E |
| 21 | 4.59 | 60 | 2.667 | 0.76 | 180 | 45.6 | 0 | 601 | 1.29 | C88 | 6.68E-06 |
| 22 | 4.59 | 60 | 2.667 | 0.65 | 180 | 93.4 | 0 | 601 | 1.29 | C88 | 8.79E-06 |
| 23 | 4.59 | 60 | 2.667 | 0.63 | 180 | 121.7 | 0 | 601 | 1.29 | C88 | 1.18E-05 |
| 24 | 4.59 | 60 | 2.667 | 0.53 | 180 | 257.8 | 0 | 601 | 1.29 | C88 | 2.15E-05 |
| 25 | 4.59 | 30 | 2.667 | 0.64 | 180 | 195.6 | 0 | 196 | 0 | SS 316L | 4.55E-05 |
| 26 | 4.59 | 40 | 2.667 | 0.64 | 180 | 195.6 | 0 | 196 | 0 | SS 316L | 5.31E-05 |
| 27 | 4.59 | 50 | 2.667 | 0.64 | 180 | 195.6 | 0 | 196 | 0 | SS 316L | 6.34E-05 |
| 28 | 4.59 | 30 | 2.667 | 0.64 | 180 | 195.6 | 0 | 601 | 1.29 | C88 | 1.36E-05 |
| 29 | 4.59 | 40 | 2.667 | 0.64 | 180 | 195.6 | 0 | 601 | 1.29 | C88 | 1.69E-05 |
| 30 | 4.59 | 50 | 2.667 | 0.64 | 180 | 195.6 | 0 | 601 | 1.29 | C88 | 1.83E-05 |

*Source:* PhD Thesis; Singh [26].

$V$, velocity (m/s); $C$, concentration (% weight); $k$, bulk density; $CF$, circularity factor of erodent; $T$, time (min); $d_p$, particle size (μm); $A$, impact angle (°); $H$, microhardness (HV); $P$, porosity; C88, Colmonoy 88 HVOF coating, high-velocity oxy-fuel coating.

present study. This dataset is converted row-wise for an accurate reading. However, the output file should contain the same rows as the input file for an accurate reading of the data. Against the $9 \times 30$ matrix data, the output file used in the present study was $1 \times 30$ matrix. The output was the actual experimental data from erosion experiments. Regression models were trained using the MATLAB tool with the input parameters and an output parameter, i.e., erosion rate. In this study, the training, validation, and testing data were selected at 75%, 15%, and 15%, respectively.

## 8.6   RESULTS FROM REGRESSION MODELS

In the present study, the evaluation of the prepared SVM model was done on the basis of the root mean square error (RMSE) and Pearson correlation coefficient ($R$). The designed SVM model was compared with other popular regression models such as GPR, Linear Regression (LR), Ensemble Boost (EnBoost), and Coarse Tree, as shown in Figure 8.4.

The values that are predicted by the model have a high degree of correspondence with the values that are discovered in real experiments. The error histogram that was generated by the SVM when it was being trained can be found in Figure 8.5. The performance of the ANN model is evaluated based on an error percentage that falls between 0% and 6%. Figure 8.6 displays the Pearson coefficient ($R$) [28–30] for the training, validation, and testing phases. The Pearson correlation coefficient ($R$) [31–34] is given by:

$$R = \frac{A \sum pq - \left( \sum p \sum q \right)}{\sqrt{\left[ A \sum p^2 - \left( \sum p \right)^2 \right]\left[ A \sum q^2 - \left( \sum q \right)^2 \right]}} \qquad (8.5)$$

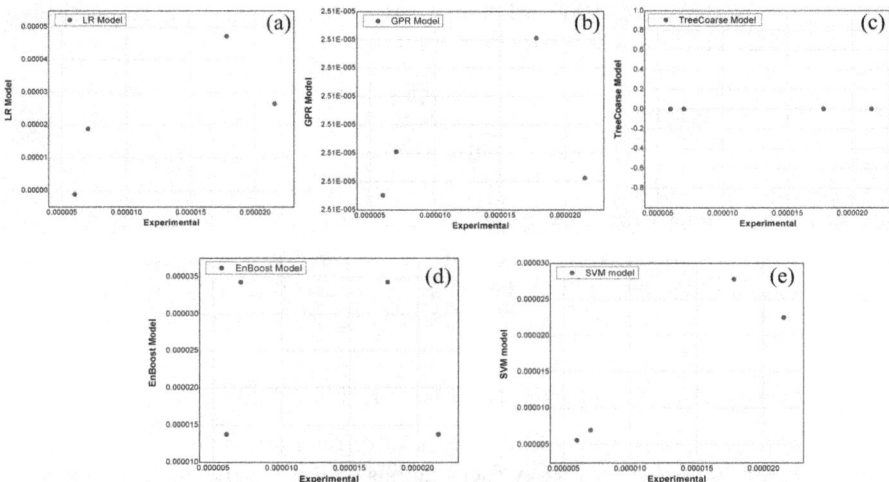

FIGURE 8.4   (a) Linear regression, (b) Gaussian process regression, (c) TreeCoarse, (d) ensemble Boost, and (e) support vector machine model.

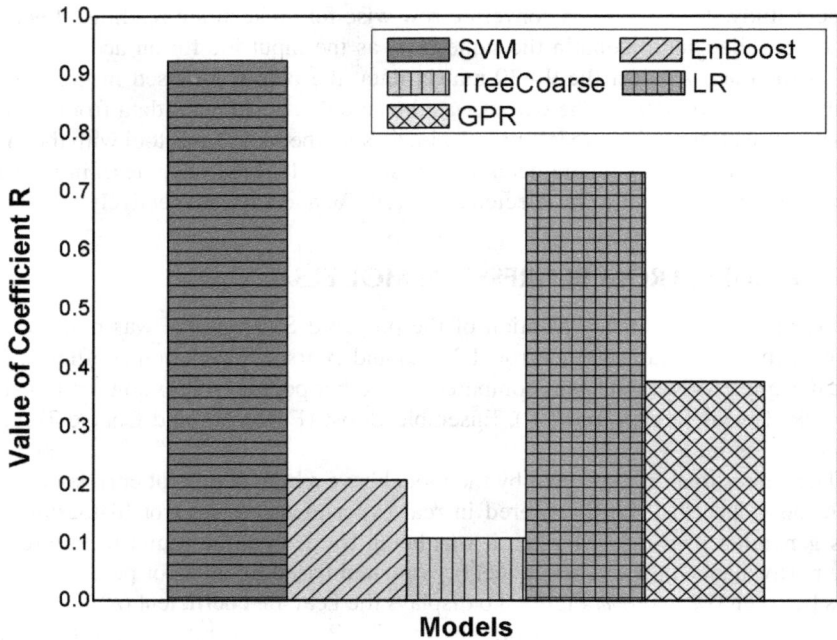

**FIGURE 8.5** *R*-value of the regression models.

where $A$ is the pair scores. $\sum pq$ is the product of paired scores. Symbols $\sum p$ and $\sum p$ are the SS (sum of scores). The $R$-value of regression models is shown in Figure 8.5. The values of the Pearson coefficient ($R$) were calculated as 0.924, 0.21, 0.11, 0.732, and 0.377 for SVM, EnBoost, TreeCoarse, LR, and GPR, respectively.

RMSE [35] was also calculated as a performance measure:

$$\text{RMSE} = \sqrt{\frac{1}{x} \sum_{n=1}^{k} (r_n - \bar{r}_n)^2} \qquad (8.6)$$

where $x$ is the measure of the total observations. The $r_n$ and $\bar{r}_n$ are the predicted and observed values, respectively. The RMSE value of regression models is shown in Figure 8.6. The RMSE was found to be $5.03 \times 10^{-6}$, $1.69 \times 10^{-5}$, $1.38 \times 10^{-5}$, $1.64 \times 10^{-5}$, and $1.38 \times 10^{-5}$ for SVM, EnBoost, TreeCoarse, LR, and GPR, respectively.

## 8.7 CONCLUSIONS AND FUTURE PERSPECTIVE

In this chapter, a neural network prediction of erosion rate in stainless steel 316L and Colmonoy 88 high-velocity oxy-fuel coatings was carried out. The results of this case study indicate that the newly developed SVM model has improved prediction ability. This error range of 0–8% demonstrates the effectiveness of the SVM model.

# Index

[27] Ortegon J, Ledesma-Alonso R, Barbosa R, et al. Material phase classification by means of support vector machines. *Comput Mater Sci* [Internet]. 2018;148:336–342. Available from: https://doi.org/10.1016/j.commatsci.2018.02.054.

[28] Qian N. On the momentum term in gradient descent learning algorithms. *Neural Networks*. 1999;12:145–151.

[29] Hagan MT, Menhaj M. Training feedforward networks with the Marquardt algorithm. *IEEE Trans Neural Netw*. 1994;5:989–993.

[30] Levenberg K. A method for the solution of certain non-linear problems in least squares. *Q Appl Math*. 1944;2:164–168.

[31] Singh J. Tribo-performance analysis of HVOF sprayed 86WC-10Co4Cr & Ni-Cr2O3 on AISI 316L steel using DOE-ANN methodology. *Ind Lubr Tribol*. 2021;73:727–735.

[32] Benesty J, Chen J, Huang Y, et al. *Pearson Correlation Coefficient. Noise Reduct Speech Process Springer Top Signal Process*, Vol. 2. Berlin, Heidelberg: Springer-Verlag; 2009. pp. 37–40.

[33] Kumar P, Singh J, Singh S. Neural network supported flow characteristics analysis of heavy sour crude oil emulsified by ecofriendly bio-surfactant utilized as a replacement of sweet crude oil. *Chem Eng J Adv* [Internet]. 2022;11:100342. Available from: https://doi.org/10.1016/j.ceja.2022.100342.

[34] Singh J, Singh S. Neural network supported study on erosive wear performance analysis of Y2O3/WC-10Co4Cr HVOF coating. *J King Saud Univ - Eng Sci* [Internet]. 2022; in press. Available from: https://doi.org/10.1016/j.jksues.2021.12.005.

[35] Singh J, Singh S. A review on machine learning aspect in physics and mechanics of glasses. *Mater Sci Eng B* [Internet]. 2022;284:115858. Available from: https://doi.org/10.1016/j.mseb.2022.115858.

[8] Raissi M, Perdikaris P, Karniadakis GE. Machine learning of linear differential equations using Gaussian processes. *J Comput Phys* [Internet]. 2017;348:683–693. Available from: https://doi.org/10.1016/j.jcp.2017.07.050.

[9] Efromovich S. *The Multivarient Normal Distribution*. New York: Springer Series in Statistics, Springer-Verlag; 1999.

[10] Bishnoi S, Singh S, Ravinder R, et al. Predicting Young's modulus of oxide glasses with sparse datasets using machine learning. *J Non Cryst Solids* [Internet]. 2019;524:119643. Available from: https://doi.org/10.1016/j.jnoncrysol.2019.119643.

[11] Rustam Z, Yaurita F. Insolvency prediction in insurance companies using support vector machines and fuzzy kernel C-means. *J Phys Conf Ser*. 2018;1028:012118.

[12] Villarrubia G, De Paz JF, Chamoso P, et al. Artificial neural networks used in optimization problems. *Neurocomputing*. 2018;272:10–16.

[13] Raj JS, Ananthi JV. Recurrent neural networks and LSTM explained. *J Soft Comput Paradig* [Internet]. 2019;01:33–40. Available from: https://medium.com/@purnasaigudikandula/recurrent-neural-networks-and-lstm-explained-7f51c7f6bbb9.

[14] Deng B. Machine learning on density and elastic property of oxide glasses driven by large dataset. *J Non Cryst Solids* [Internet]. 2020;529:119768. Available from: https://doi.org/10.1016/j.jnoncrysol.2019.119768.

[15] Friedman JH. Greedy functional approximation: A gradient boosting machine. *Ann Stat* [Internet]. 2001;29:1189–1232. Available from: https://marefateadyan.nashriyat.ir/node/150.

[16] Pomerantsev AL. Confidence intervals for nonlinear regression extrapolation. *Chemom Intell Lab Syst*. 1999;49:41–48.

[17] Wei Y, Yang F, Wainwright MJ. Early stopping for kernel boosting algorithms: A general analysis with localized complexities. *IEEE Trans Inf Theory*. 2019;65:6685–6703.

[18] Cittanti D, Ferraris A, Airale A, et al. Modeling Li-ion batteries for automotive application: A trade-off between accuracy and complexity. *Int Conf Electr Electron Technol Automotive*, Turin, Italy; 2017. pp. 1–8.

[19] Ostertagová E. Modelling using polynomial regression. *Procedia Eng* [Internet]. 2012;48:500–506. Available from: https://doi.org/10.1016/j.proeng.2012.09.545.

[20] Wang Z, Joshi S, Savel'Ev S, et al. Fully memristive neural networks for pattern classification with unsupervised learning. *Nat Electron* [Internet]. 2018;1:137–145. Available from: https://doi.org/10.1038/s41928-018-0023-2.

[21] Gholami R, Fakhari N. Support vector machine: Principles, parameters, and applications. In: Samui P, Sekhar S, Balas VE, editors, *Handbook of Neural Computation* [Internet]. 1st ed. Amsterdam: Elsevier Inc.; 2017. pp. 515–535. Available from: https://doi.org/10.1016/B978-0-12-811318-9.00027-2.

[22] Liu Y, Zhao T, Ju W, et al. Materials discovery and design using machine learning. *J Mater* [Internet]. 2017;3:159–177. Available from: https://doi.org/10.1016/j.jmat.2017.08.002.

[23] Liu L, Huang W, Wang C. Hyperspectral image classification with kernel-based least-squares support vector machines in sum space. *IEEE J Sel Top Appl Earth Obs Remote Sens*. 2018;11:1144–1157.

[24] Liu Y, Wu J, Yang G, et al. Predicting the onset temperature (Tg) of GexSe1−x glass transition: A feature selection based two-stage support vector regression method. *Sci Bull* [Internet]. 2019;64:1195–1203. Available from: https://doi.org/10.1016/j.scib.2019.06.026.

[25] Bai C, Huang L, Pan X, et al. Optimization of deep convolutional neural network for large scale image retrieval. *Neurocomputing* [Internet]. 2018;303:60–67. Available from: https://doi.org/10.1016/j.neucom.2018.04.034.

[26] Singh J. *Investigation on Slurry Erosion of Different Pumping Materials and Coatings*. Patiala, India: Thapar Institute of Engineering and Technology; 2019.

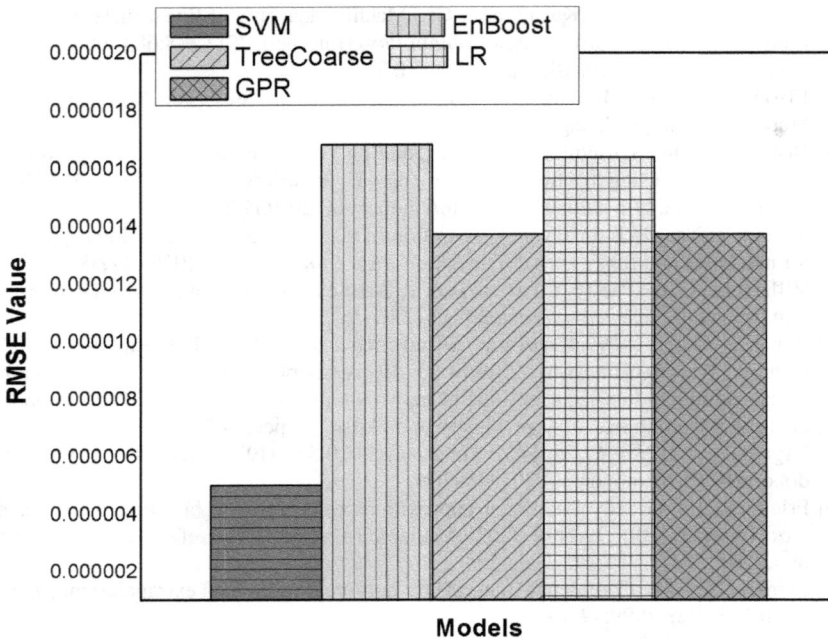

**FIGURE 8.6**    Root mean square error of the regression model.

Pearson coefficient and RMSE readings show that the SVM model prediction error was very low as compared to EnBoost, TreeCoarse, LR, and GPR. The machine learning regression tools are very immersive in terms of predicting erosion wear. In the future, the hyper-tuning of parameters can be performed to improve the prediction results from regression tools.

## REFERENCES

[1] Hunter EA, Brook RJ, Arnold GC. Applied regression analysis and experimental design. *Appl Stat.* 1986;35:77.

[2] Montgomery DC,Peck EA, Geoffrey Vining G. *Introduction to Linear Regression Analysis.* 5th ed. Wiley Series in Probability and Statistics, Hoboken: Wiley 2012.

[3] Nielson GM. Some piecewise polynomial alternatives to splines under tension [Internet]. In: Barnhill RE, Riesenfeld RF, editors. *Computer Aided Geometric Design.* Cambridge, Massachusetts: Academic Press, Inc.; 1974, pp. 209-235. Available from: https://doi.org/10.1016/B978-0-12-079050-0.50015-1.

[4] Motulsky HJ, Ransnas LA. Fitting curves nonlinear regression: Review a practical. *FASEB J.* 1987;1:365–374.

[5] Čížek P, Sadıkoğlu S. Robust nonparametric regression: A review. *Wiley Interdiscip Rev Comput Stat.* 2020;12:1–16.

[6] Altman NS. An introduction to kernel and nearest-neighbor nonparametric regression. *Am Stat.* 1992;46:175–185.

[7] Fan D, Zhang X. Short-term traffic flow prediction method based on balanced binary tree and K-nearest neighbor nonparametric regression. *Adv Intell Syst Res.* 2017;132:118–121.

For Product Safety Concerns and Information please contact our EU
representative GPSR@taylorandfrancis.com
Taylor & Francis Verlag GmbH, Kaufingerstraße 24, 80331 München, Germany